Mechanisms and Machine Science

Volume 56

Series editor

Marco Ceccarelli
LARM: Laboratory of Robotics and Mechatronics
DICeM: University of Cassino and South Latium
Via Di Biasio 43, 03043 Cassino (Fr), Italy
e-mail: ceccarelli@unicas.it

More information about this series at http://www.springer.com/series/8779

Vigen Arakelian
Editor

Dynamic Decoupling of Robot Manipulators

Springer

Editor
Vigen Arakelian
Génie Mécanique et Automatique
Institut National des Sciences Appliquées
Rennes
France

ISSN 2211-0984 ISSN 2211-0992 (electronic)
Mechanisms and Machine Science
ISBN 978-3-030-08973-3 ISBN 978-3-319-74363-9 (eBook)
https://doi.org/10.1007/978-3-319-74363-9

Softcover re-print of the Hardcover 1st edition 2018

Printed on acid-free paper

This Springer imprint is published by the registered company Springer International Publishing AG part of Springer Nature
The registered company address is: Gewerbestrasse 11, 6330 Cham, Switzerland

Preface

A manipulator design theory for reduced dynamic complexity is well known. The kinematic structure and mass distribution of a manipulator allow one to obtain a linear system, i.e. single-input and single-output system with constant parameters. It simplifies the optimal control and accumulation of energy in manipulators. There are two ways to create the dynamically decoupled manipulators: via optimal mechanical design or control.

This work presents new research results in the field of dynamic decoupling of robot manipulators, as well as some decoupled parallel structures, which have not been sufficiently disclosed in the technical literature. It includes the research results obtained in France, Russia, China and Austria.

In this book, a special place is given to the mechatronic solutions. They will certainly improve the known design concepts permitting the dynamic decoupling of serial manipulators with a relatively small increase in total mass of the moving links, and they take into account the changing payload. For the first time, such an approach has been applied on serial manipulators.

Kinematic decoupling leading to the dynamic decoupling has also been considered. Various original structures are proposed and studied in the book.

Also of great interest is the dynamic decoupling control of parallel manipulators. Firstly, the dynamic model of redundant multi-axial vibration table with load has been established, and its dynamic coupling characteristics have been analysed. Experimental results indicated that the modal space decoupling control can reduce the coupling between degrees of freedom.

The editor thanks the authors who have contributed with very interesting research results on several subjects of dynamic decoupling of robot manipulators. He hopes that the present book will be useful to the readers and it will expand knowledge in the field of robot dynamics.

Rennes, France
October 2017

Vigen Arakelian

Preface

Contents

Chapter 1
Dynamic Decoupling of Robot Manipulators: A Review with New Examples

V. Arakelian, J. Xu and J. P. Le Baron

Abstract In this chapter, a review of the main methods permitting to achieve the dynamic decoupling of robot manipulators is presented. The design approaches based on the variation of mechanical parameters are disclosed via three sub-groups: decoupling of dynamic equations via mass redistribution; decoupling of dynamic equations via actuator relocation and decoupling of dynamic equations via addition of auxiliary links. The last approach is illustrated via two examples. In the first solution, the optimal design is achieved via gears used as counterweights. It is allows a considerable reduction of the total masses of links of the decoupled manipulator. In the second solution, the dynamic decoupling of robot manipulators is achieved by using an epicyclic gear train. Special attention is paid to the dynamic decoupling of robot manipulators through the use of the double integrator. The second-order linear and time-invariant dynamical system, called double integrator, is one of the most fundamental systems in control applications. It can be considered as single-degree-of-freedom translational and rotational motion. The present review considers in detail the aim of this solution, as well as the advantages of the joint application development inclosing mechanical and control solutions.

1.1 Dynamics and Control of Manipulators

It is known that the manipulator dynamics are highly coupled and nonlinear. The complicated dynamics results from varying inertia, interactions between the different joints, and nonlinear forces such as Coriolis and centrifugal forces. The aim of dynamic decoupling of manipulators is to ensure the conditions which allow one to obtain decoupled and linear dynamic equations. It simplifies the optimal control

V. Arakelian (✉) · J. Xu · J. P. Le Baron
Department of Mechanical and Control Systems Engineering, INSA,
20 avenue des Buttes de Coësmes, CS 70839, 35708 Rennes Cedex, France
e-mail: vigen.arakelyan@insa-rennes.fr

V. Arakelian · J. P. Le Baron
LS2 N/ECN, 1 rue de la Noë, BP 92101, 44321 Nantes Cedex 03, France

V. Arakelian (ed.), *Dynamic Decoupling of Robot Manipulators*, Mechanisms and Machine Science 56, https://doi.org/10.1007/978-3-319-74363-9_1

and accumulation of energy in manipulators. There are two ways to create the dynamically decoupled manipulators: via optimal mechanical design or control.

Let us firstly consider the dynamics and control of serial manipulators in order to show the dynamic decoupling by using the control of the double integrator.

An effective way to deal with the problem of high complex coupled dynamics is the decoupling. In fact, this concept was first proposed by Morgan in 1964 when he tried to introduce the design method of the typical control theory into the MIMO (Multi-Input and Multi-Output) linear system [1]. Morgan searches the necessary condition in order that the closed-loop transfer function matrix is a full rank diagonal rational matrix.

As known, the form of the dynamic equation of the serial manipulator can be written as:

$$\tau = M(\theta)\ddot{\theta} + C(\theta, \dot{\theta})\dot{\theta} + g(\theta) \tag{1.1}$$

where τ is a $n \times 1$ torque vector applied to the joints of the manipulator; $\theta, \dot{\theta}$ and $\ddot{\theta}$ are $n \times 1$ vectors representing the angular positions, velocities and accelerations, respectively; $M(\theta)$ is $n \times n$ inertia matrix; $C(\theta, \dot{\theta})$ is $n \times 1$ vector of Coriolis and centrifugal effects; $g(\theta)$ is the torque vector due to gravity.

The first term on the right hand side represents the inertia torque. It can be divided into two parts: the main diagonal elements of the matrix and the off-diagonal elements of the matrix. The first part is generated by the acceleration of the corresponding joint.

The second part is the interactive inertia torque caused by the accelerations of the other joints. This interactive inertia torque is linearly proportional to acceleration. The second term represents the nonlinear velocity torques resulting from Coriolis and centrifugal effects. Generally, the dependence of the inertia matrix on the arm configuration produces these nonlinear velocity torques.

For an arbitrary arm configuration, the inertia matrix $M(\theta)$ is reduced to a diagonal matrix if the off-diagonal elements of the matrix $M(\theta)$ are all zero. Then it is defined as decoupled inertia matrix. Hence, the control system can be treated as a set of SISO (Single-Input and Single-Output) subsystems.

When the second term $C(\theta, \dot{\theta})\dot{\theta}$ disappears, the inertia matrix is constant for all arbitrary arm configurations (the matrix $M(\theta) = M$ is independent of joint displacements). In this case, the inertia matrix is referred to as configuration invariant inertia matrix. The significance of this form is that the linear control methods which are much simpler and easier to implement can be adopted.

However, the most desirable form for the manipulator dynamics is the one with decoupled and configuration-invariant inertia matrix where the effects of gravity are compensated by mechanical engineering, that is

$$\tau = M\ddot{\theta} \tag{1.2}$$

where M is $n \times n$ diagonal inertia matrix.

In this case, the system is completely decoupled and linearized. This system can be treated as SISO systems with constant parameters.

As the main inherent character of the serial manipulators, the dynamic coupled results from varying inertia, interactions between the different joints and nonlinear forces such as Coriolis and centrifugal forces have obvious influence on the positioning accuracy. This influence exists both at slow speed and high speed, especially for the latter. Hence, many researchers devote to achieve the dynamic decoupling and linear situation. When this situation is realized:

- dynamic equation can be simplified and the controller can be treated as the superposition of serial SISO controller;
- the controller design can be simplified, hence the real-time operational performance of the manipulator can be improved because of the reduction of the computational burden;
- positioning error can be reduced by the eliminations of the Coriolis and centrifugal forces. Therefore the robustness of the serial manipulators is improved.

In the part of control, lots methods proposed out to reduce or even to eliminate such dynamic complexity of the serial manipulators. Let us consider about some aspects of the control: (a) PD control of serial manipulators; (b) inverse dynamics control of serial manipulators; (c) control of the double integrator.

(a) PD control of serial manipulators

The control of most industrial manipulators in use today is based on the application of conventional servo control techniques such as PD (proportional-derivative) or PID (proportional-integral-derivative) control [2–8]. In this kind of method, the tracking errors which describe the difference between the desired and real trajectories are multiplied by gains, and then contribute as part of input torque to reduce the difference. This traditional control method is widely used because of its simple structure, easy implementation and robust operation.

It has been shown that a simple PD control applied at each joint is adequate in most position control applications such as spot-welding or palletizing. An independent joint PD controller has been shown to be asymptotic stable for rigid manipulators [9]. Consider an equation, without the effects of friction, where the gravity is compensated by the control law:

$$M(\theta)\ddot{\theta} + C(\theta, \dot{\theta})\dot{\theta} = \tau - g(\theta) = u \tag{1.3}$$

where

$$u = K_P(\theta_d - \theta) - K_D\dot{\theta} \tag{1.4}$$

Be an independent joint PD control, where θ_d represents a constant reference set-point; K_P and K_D are positive, diagonal matrices of proportional and derivative gains, respectively. Consider the Lyapunov function candidate

$$V = \frac{1}{2}\dot{\theta}^T M(\theta)\dot{\theta} + \frac{1}{2}(\theta_d - \theta)^T K_P(\theta_d - \theta) > 0 \tag{1.5}$$

Then a simple calculation using the skew symmetry property shows that:

$$\dot{V} = -\dot{\theta}^T K_D \dot{\theta} \leq 0 \tag{1.6}$$

The function $\dot{V}$ is not negative definite. In this case, the LaSalle's theorem [10], known today as LaSalle's invariance principle, may be used to prove the asymptotic stability.

$$\dot{\theta} = 0 \Rightarrow \ddot{\theta} = 0 \Rightarrow M(\theta)\ddot{\theta} + C(\theta, \dot{\theta})\dot{\theta} = K_P(\theta_d - \theta) - K_D\dot{\theta} = 0 \Rightarrow \theta = \theta_d \tag{1.7}$$

However, as shown here, the gravity parameters must be known exactly. The other limitation is simply due to the inherent "mismatch" between the nonlinear dynamics character of the manipulators and the linear regulating behavior of the PID or PD controller [11]. Hence, this control method is not satisfactory for applications which require high tracking accuracy and high speed performance.

(b) Inverse dynamics control of serial manipulators

In inverse dynamics control method, the system outputs are assumed as anticipative values. Then the corresponding inputs are calculated through the reversed state function. So, the inputs are functions of time and state variables. This is an inverse system of the original system. Finally, the two systems combine together compose an artificial linear system. Based on time-scale separation principle, the system can be separated into an inner control loop subsystem and an outer control loop subsystem by time-scale [12]. If the inner control loop has already realized decoupling and obtained control performance well, the outer control loop can be simplified with classical control method. So the crux of inverse dynamics control is the design of the inner loop (Fig. 1.1).

Consider the dynamic model of a manipulator that described by Eq. (1.1), then the nonlinear feedback control law is given as [13]:

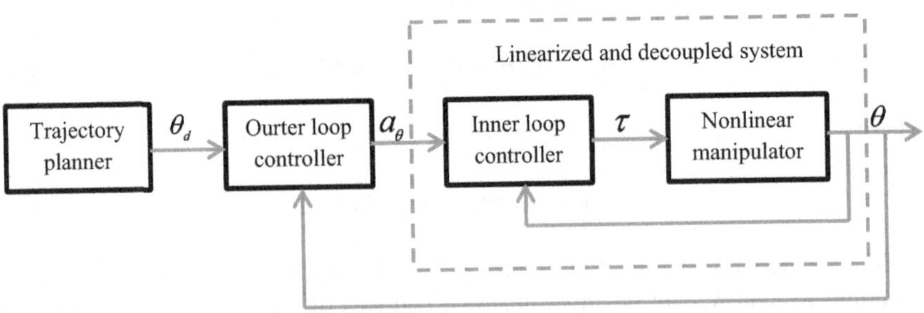

Fig. 1.1 The control scheme of inverse dynamics control

$$\tau = M(\theta)a_\theta + C(\theta,\dot{\theta})\dot{\theta} + g(\theta) \tag{1.8}$$

where $a_\theta \in \mathbb{R}^n$ is, as yet, undetermined. Since the inertia matrix $M(\theta)$ is invertible for all θ, the closed-loop system reduces to the decoupled double integrator.

$$\ddot{\theta} = a_\theta \tag{1.9}$$

Given a joint space trajectory, $\theta_d(t)$, an obvious choice for the outer loop term a_θ is as a PD plus feedforward acceleration control

$$a_\theta = \ddot{\theta}_d + K_P(\theta_d - \theta) - K_D(\dot{\theta}_d - \dot{\theta}) \tag{1.10}$$

Substituting Eq. (1.10) into Eq. (1.9) and defining

$$\tilde{\theta} = \theta - \theta_d \tag{1.11}$$

The linear and decoupled closed-loop system is obtained as

$$\ddot{\tilde{\theta}} + K_D\dot{\tilde{\theta}} + K_P\tilde{\theta} = 0 \tag{1.12}$$

In the robotics context, feedback linearization is also known as inverse dynamics. Although this method is possible in theory, it is difficult to achieve in practice, mainly because the coordinate transformation is a function of the system parameters and, hence, sensitive to uncertainty.

(c) Control of the double integrator

The second-order linear and time-invariant dynamical system, called double integrator, is one of the most fundamental systems in control applications. It can be considered as single-degree-of-freedom translational and rotational motion. So, researchers are interested in double integrator since the early days of control theory when it was used extensively to illustrate minimum-time and minimum-fuel controllers [14–16]. The equations of the double integrator are given by [17].

$$\begin{aligned} \dot{x} &= Ax + Bu \\ y &= Cx \end{aligned} \tag{1.13}$$

where

$$x = \begin{bmatrix} \theta \\ \dot{\theta} \end{bmatrix}, A = \begin{bmatrix} 0 & 1 \\ 0 & 0 \end{bmatrix}, B = \begin{bmatrix} 0 \\ 1 \end{bmatrix}, C = [0 \quad 1].$$

Since the double integrator is unstable but completely controllable and observable, closed-loop control strategies can be used to drive the state to the origin in the finite time, some researches are carried out based on this kind of model [18]. In the other aspect, a simple linear double integrator relationship between the output and input is better for a tracking controller [19].

In this book, the design of the controller is based on the state and the control law given by $u = -Gx$.

The function of the gain matrix $G = [g_1 \quad g_2]$ is to stabilize the system by moving the closed-loop poles in the left-half complex plane. We seek u that minimizes the cost

$$J = \int_0^\infty \left[Ly^2 + u^2\right] dt = \int_0^\infty \left[x^T Q_C x + u^2\right] dt \tag{1.14}$$

The matrix L is based on the controllability transient gramian defined by

$$G_C(0, T_P) = \int_0^{T_P} \left[e^{At} B B^T e^{A^T t}\right] dt \tag{1.15}$$

For the matrix $L = \left[T_P C G_C(0, T_P) C^T\right]^{-1}$, the matrix $Q_C = C^T L C$ is symmetric and semi-definite positive. The parameter T_p assume that poles of closed-loop system may be placed, in the S plane, at the left or near of the vertical straight with the abscissa $-\frac{1}{T_P}$. The output equation $u = -Gx$ of the controller is unique, optimal, full state feedback control law with $G = B^T \Sigma_C$ that minimizes the cost J.

The matrix Σ_C is the unique, symmetric, positive definite solution to the algebraic Riccati equation $A^T \Sigma_C + \Sigma_C A - \Sigma_C B B^T \Sigma_C + Q_C = 0$.

For the double integrator, the matrix G gives: $G = \left[g_1 = \frac{\sqrt{3}}{T_P^2} \quad g_2 = \frac{\sqrt{2\sqrt{3}}}{T_P} \right]$.

Then the closed-loop characteristic polynomial is: $P_C(s) = s^2 + \frac{\sqrt{2\sqrt{3}}}{T_P} s + \frac{\sqrt{3}}{T_P^2}$.

If $P_C(s) = s^2 + 2\zeta\omega_n s + \omega_n^2$, we have: $\omega_n = \frac{\sqrt{\sqrt{3}}}{T_P}$ and $\zeta = \frac{\sqrt{2}}{2}$.

The closed-loop control law, presented by Fig. 1.2, can be written as

$$u = \ddot{\theta}_R - g_1[\theta - \theta_R] - g_2[\dot{\theta} - \dot{\theta}_R] \tag{1.16}$$

In the case of some characterization of the model uncertainties, we propose a robust controller based from estimate state and implicit integral action [20].

The zero-steady-state error optimal control law is given by $u(t) = -G\hat{x} - \hat{d}$.

The gain matrix G is given by same manner as previously because regulator and observer can be design separately without destroying stability properties for each other. The estimate of the state-vector x is $\hat{x}$ and $\hat{d}$ is an estimate of the constant disturbance d placed in the state equation of the double integrator.

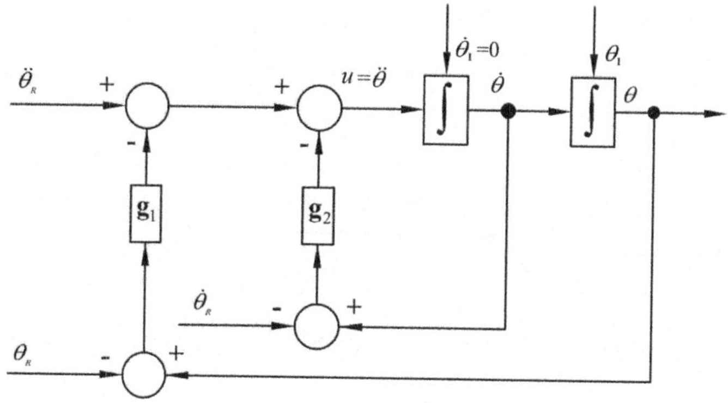

Fig. 1.2 The closed-loop control law of double integrator system

For obtain the observer, we model the constant disturbance as $\dot{d} = 0$. The steady-state optimal observer which allows estimating x and d is

$$\underbrace{\begin{bmatrix} \dot{\hat{\theta}} \\ \ddot{\hat{\theta}} \end{bmatrix}}_{\dot{\hat{x}}} = \underbrace{\begin{bmatrix} 0 & 1 \\ 0 & 0 \end{bmatrix}}_{A} \underbrace{\begin{bmatrix} \hat{\theta} \\ \dot{\hat{\theta}} \end{bmatrix}}_{\hat{x}} + \underbrace{\begin{bmatrix} 0 \\ 1 \end{bmatrix}}_{B} u(t) + \underbrace{\begin{bmatrix} 0 \\ 1 \end{bmatrix}}_{B} \hat{d} + \begin{bmatrix} k_1 \\ k_2 \end{bmatrix} \left(y - \underbrace{\begin{bmatrix} 1 & 0 \end{bmatrix}}_{C} \underbrace{\begin{bmatrix} \hat{\theta} \\ \dot{\hat{\theta}} \end{bmatrix}}_{\hat{x}} \right) \tag{1.17}$$

$$\dot{\hat{d}}(t) = k_3 \left(y - \underbrace{\begin{bmatrix} 1 & 0 \end{bmatrix}}_{C} \underbrace{\begin{bmatrix} \hat{\theta} \\ \dot{\hat{\theta}} \end{bmatrix}}_{\hat{x}} \right) \tag{1.18}$$

The state-equations of the observer are:

$$\underbrace{\begin{bmatrix} \dot{\hat{\theta}} \\ \ddot{\hat{\theta}} \\ \dot{\hat{d}} \end{bmatrix}}_{\dot{x}_E} = \underbrace{\begin{bmatrix} -k_1 & 1 & 0 \\ -k_2 & 0 & 1 \\ -k_3 & 0 & 0 \end{bmatrix}}_{A_E} \underbrace{\begin{bmatrix} \hat{\theta} \\ \dot{\hat{\theta}} \\ \hat{d} \end{bmatrix}}_{x_E} + \underbrace{\begin{bmatrix} 0 \\ 1 \\ 0 \end{bmatrix}}_{B_E} u + \underbrace{\begin{bmatrix} k_1 \\ k_2 \\ k_3 \end{bmatrix}}_{K} y \tag{1.19}$$

where, $y = \underbrace{\begin{bmatrix} 1 & 0 & 0 \end{bmatrix}}_{C_E} x_E$.

The function of the gain matrix $K = \begin{bmatrix} k_1 & k_2 & k_3 \end{bmatrix}^T$ is to stabilize asymptotically the observer. The duality between the optimal regulator and the optimal

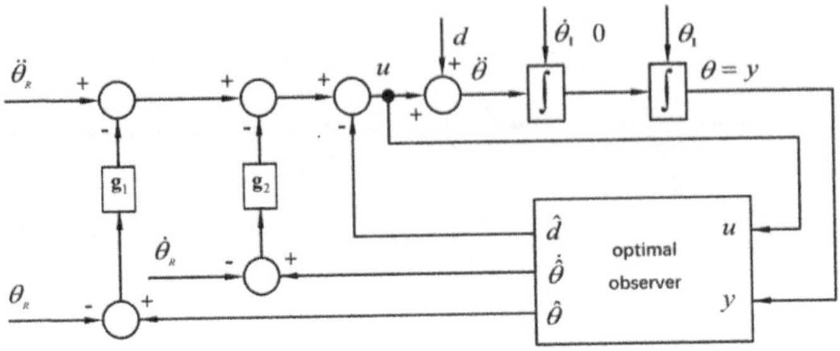

Fig. 1.3 The closed-loop control system of the double integrator with optimal observer block

observer (Kalman filter) enables us to transfer from the regulator to the observer all important results. The Riccati equation can be rephrased as $A_E\Sigma_O + \Sigma_O A_E^T - \Sigma_O C_E^T C_E \Sigma_O + Q_O = 0$.

The matrix $Q_O = [T_R G_O(0, T_R)]^{-1}$ is based on the observability transient gramian defined by $G_O(0, T_R) = \int_0^{T_R} \left[e^{A_E^T t} C_E^T C_E e^{A_E t} \right] dt$. The parameter T_R assumes that poles of the observer may be placed, in the s plane, at the left or near of the vertical straight with the abscissa $-\frac{1}{T_R}$. The solution of the Riccati equation is given by:

$$K = \Sigma_O C_E^T = \left[k_1 = \frac{c_1}{T_R} \quad k_2 = \frac{c_2}{T_R^2} \quad k_3 = \frac{c_3}{T_R^3} \right]^T \tag{1.20}$$

For $c_1^2 - 2c_2 = 9$ and $c_3 = 12\sqrt{5}$, the values are: $c_1 = 7.198$; $c_2 = 21.408$; $c_3 = 26.83$.

Then the characteristic polynomial is:

$$P_O(s) = \left(s + \frac{3.0735}{T_R} \right) \left(s^2 + \frac{4.1248}{T_R} s + \frac{8.7303}{T_R^2} \right).$$

If $P_O(s) = (s + \omega_1)(s^2 + 2\zeta\omega_0 s + \omega_0^2)$, we have $\omega_0 = \frac{2.9547}{T_R}$ and $\zeta = 0.698$.

The Fig. 1.3 shows the closed-loop control system which accumulates information about the double integrator during operation and allows a zero steady-state tracking error in spite of constant disturbance d defined by $\dot{d} = 0$.

The closed-loop control law can be written as

$$u = \ddot{\theta}_R - g_1 \left[\hat{\theta} - \theta_R \right] - g_2 \left[\hat{\dot{\theta}} - \dot{\theta}_R \right] - \hat{d} \tag{1.21}$$

1.2 Design of Manipulators with Linear and Decoupled Dynamics

As mentioned above, the dynamics of a manipulator arm depends upon the mass properties of individual arm links and the kinematic structure of the arm linkages. Hence, usually, the redistribution of mass and modification of the arm structure are used to reducing the dynamic complexity of the manipulators.

Let us consider three main approaches developed for dynamic decoupling of manipulators via mechanical transformation: (d) via mass redistribution; (e) via actuator relocation and (f) via addition of auxiliary links.

(d) Decoupling of dynamic equations via mass redistribution

The necessary conditions for the decoupled and configuration-invariant inertia of the general manipulator are given in [21]. Let us disclose the necessary properties of mass distribution to reach this goal.

Let θ_i and τ_i, be the joint displacement and torque of the i-th joint, respectively, then the equation of motion of the manipulator is given by

$$\tau_i = H_{ii}\ddot{\theta}_i + \sum_{j \neq i} H_{ij}\ddot{\theta}_j + \sum_j \sum_k \left(\frac{\partial H_{ij}}{\partial \theta_k} - \frac{1}{2}\frac{\partial H_{jk}}{\partial \theta_i} \right) \dot{\theta}_j \dot{\theta}_k + \tau_{gi} \tag{1.22}$$

where H_{ij} is the i-j element of the manipulator inertia matrix, and τ_{gi} is the torque due to gravity. The first term of this equation represents the inertia torque generated by the acceleration of the ith joint, while the second term is the interactive inertia torque caused by the accelerations of the other joints. The interactive inertia torque is linearly proportional to acceleration. The third term represents the nonlinear velocity torques resulting from Coriolis and centrifugal effects. In general, the dependence of the inertia matrix on the arm configuration produces these nonlinear velocity torques.

Consider the inertia matrix that reduces to a diagonal matrix for an arbitrary arm configuration, then the second term in equation mentioned above vanishes and no interactive torques appear. The manipulator inertia matrix in this case is referred to as a decoupled inertia matrix. The significance of the decoupled inertia matrix is that the control system can be treated as a set of single-input, single-output subsystems associated with individual joint motions.

The equation of motion under these conditions reduces to

$$\tau_i = H_{ii}\ddot{\theta}_i + \sum_k \left(\frac{\partial H_{ii}}{\partial \theta_k} \dot{\theta}_j \dot{\theta}_k - \frac{1}{2}\frac{\partial H_{kk}}{\partial \theta_k} \dot{\theta}_k^2 \right) + \tau_{gi} \tag{1.23}$$

where the second term represents the nonlinear velocity torques resulting from the spatial dependency of the diagonal elements of the inertia matrix. Note that the number of terms involved in this equation is much smaller than the number of original nonlinear velocity torques, because all the off diagonal elements are zero for $\theta_1, \ldots, \theta_n$. This reduces the computational complexity of the nonlinear torques.

Another significant form of the inertia matrix that reduces the dynamic complexity is a configuration-invariant form. The inertia matrix in this case does not vary from an arbitrary arm configuration. In other words, the matrix is independent of joint displacements, hence the third term in first equation vanishes and the equation of motion reduces to

$$\tau_i = H_{ii}\ddot{\theta}_i + \sum_j H_{ij}\ddot{\theta}_j + \tau_{gi} \tag{1.24}$$

Note that the coefficients H_{ii} and H_{ij} are constant for all arm configurations. Thus, the equation is linear except the last term, that is, the gravity torque. The inertia matrix in this form is referred to as an invariant inertia matrix. The significance of this form is that linear control schemes can be adopted, which are much simpler and easier to implement.

When the inertia matrix is both decoupled and configuration invariant, the equation of motion reduces to

$$\tau_i = H_{ii}\ddot{\theta}_i + \tau_{gi} \tag{1.25}$$

The system is completely decoupled and linearized, except the gravity term. Thus, we can treat the system as single-input, single-output systems with constant parameters.

In order to eliminate the coupling and nonlinear torques, the inertia matrix must be diagonalized and made invariant for all the arm configurations [21].

The linearization and dynamic decoupling of 3-DOF manipulators have also been considered [22]. In this research, all of the arm constructions that yielded the decoupled inertia matrices were determined. The approach in this research is applied to serial manipulators in which the axis of joints are not parallel. In the case of parallel axes such an approach allows linearization of the dynamic equations but not their dynamic decoupling [23]. Thus, in the case of planar serial manipulators, it cannot be used.

Thus, in serial manipulator arms with an open kinematic chain structure, the inertia matrix cannot be decoupled unless the joint axes are orthogonal to each other.

(e) Decoupling of dynamic equations via actuator relocation

A popular configuration for the actuation of robot manipulators with actuated joints is to have motors directly attached to the joints. This design does not involve any transmission elements between the actuators and the joints. However, in certain

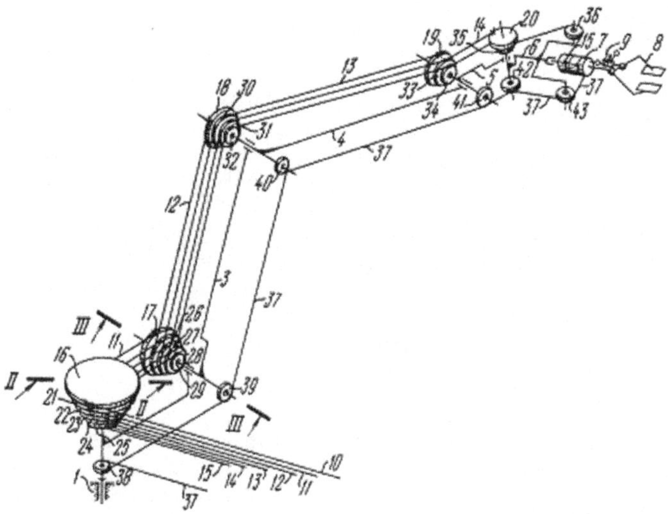

Fig. 1.4 Dynamic decoupling via actuator relocation [24]

cases this configuration may not be appropriate and manipulators with remotely-actuated joints may be desirable from point of view of the simplified dynamics.

In this case the dynamic decoupling follows from the kinematic decoupling of motion when the rotation of any link is due to only one actuator. It is obvious that it must be accompanied by an optimal choose the mass properties of certain links.

The review has shown that the design concept with remote actuation is not optimal from the point of view of the precise reproduction of the end-effector tasks because it accumulates all errors due to the clearances and elasticity of the belt transmission mainly used (Fig. 1.4).

Obviously, it is lot better to connect actuators directly with links than to use transmission mechanisms. The manufacturing and assembly errors of the added transmission mechanisms also have a negative impact to the robot precision.

(f) Decoupling of dynamic equations via addition of auxiliary links

The linearization of the dynamic equations and their decoupling via redesign of the manipulator by adding auxiliary links has also been developed [25]. The dynamic decoupling via redesign of the manipulator by adding auxiliary links is a promising new approach in robotics.

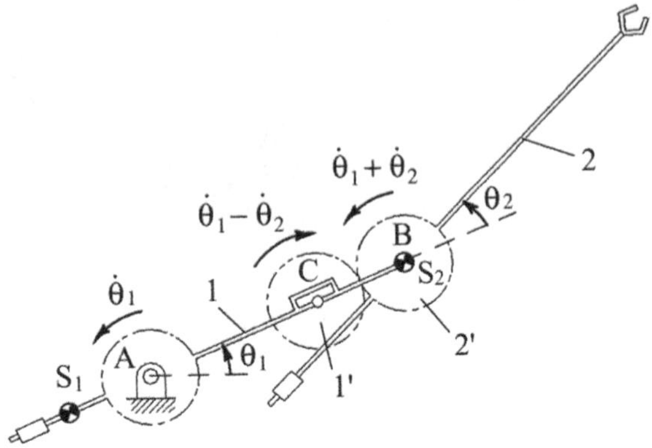

Fig. 1.5 A 2-DOF manipulator with addition of two gears

Figure 1.5 shows a modified 2-DOF serial manipulator with two added gears [25].

Ensuring condition

$$I_{S2} + I_{S2'} = I_{S1'} \tag{1.31}$$

the dynamic equations of the manipulator are decoupled:

$$\tau_1 = \left[m_1 l_{AS1}^2 + (m_2 + m_{2'}) l_1^2 + m_{1'} l_{AC}^2 + I_{S1} + I_{S1'} + I_{S2} + I_{S2'}\right] \ddot{\theta}_1 \tag{1.32}$$

$$\tau_2 = (I_{S2} + I_{S2'} + I_{S1'}) \ddot{\theta}_2 \tag{1.33}$$

where, l_1, l_2 are the lengths of links 1 and 2; l_{AS1} is the distance between the centre of mass S_1 of link 1 and joint centre A; l_{AC} is the distance between the centre of joint A and joint centre C; m_1, m_2 are the masses of links 1 and 2; I_{S1} is the axial moment of inertia of link 1 relative to the centre of mass S_1 of link 1; I_{S2} is the axial moment of inertia of link 2 relative to the centre of mass S_2 of link 2; $I_{S1'}$ is the axial moment of inertia of gear 1′; $I_{S2'}$ is the axial moment of inertia of gear 2′; $\ddot{\theta}_1$ is the angular acceleration of link 1 relative to the base; $\ddot{\theta}_2$ is the angular acceleration of link 2 relative to link 1.

However, the design methodology proposed in [25], which claims that it is the first time the added links have been used for dynamic decoupling, leads to the unavoidable increase of the total mass of the manipulator. This is due to the disposition of the added elements in the end of each link.

Let us consider a design concept which allows one to reduce the total mass of the decoupled manipulator.

1.3 Improved Dynamic Decoupling of Robot Manipulators Through the Use of Gears as Counterweights

The proposed 2-DOF manipulator illustrated in Fig. 1.6 differs from the traditional scheme by the fact that two counter rotating links 5 and 6 are mounted on actuated element 1 and linked kinematically with second actuated element 2 through a belt transmission.

Let us rewrite the Lagrangian of the manipulator with the added links:

$$
\begin{aligned}
L = K - P = \sum_{i=1}^{6} K_i - \sum_{i=1}^{6} P_i &= 0.5\left(m_1 V_{S1}^2 + m_2 V_{S2}^2 + m_3 V_B^2 + m_6 V_C^2\right) \\
&+ 0.5\left[I_{S1}\dot{\theta}_1^2 + (I_{S2} + I_{S3} + I_{S4} + I_{S5})\left(\dot{\theta}_1 + \dot{\theta}_2\right)^2 + I_{S6}\left(\dot{\theta}_1 - \dot{\theta}_2\right)^2\right] \\
&- \sum_{i=1}^{6} P_i = 0.5\left[(k_1 + 2k_2\cos\theta_2)\dot{\theta}_1^2 + k_3\dot{\theta}_2^2 + (k_4 + 2k_2\cos\theta_2)\dot{\theta}_1\dot{\theta}_2\right] \\
&- k_5\sin\theta_1 - k_6\sin(\theta_1 + \theta_2)
\end{aligned}
\tag{1.34}
$$

where,

$$
\begin{aligned}
k_1 = m_1 l_{AS1}^2 + m_2\left(l_1^2 + l_{BS2}^2\right) + m_3 l_1^2 + m_6 l_{AC}^2 + I_{S1} + I_{S2} \\
+ I_{S3} + I_{S4} + I_{S5} + I_{S6}
\end{aligned}
\tag{1.35}
$$

$$k_2 = m_2 l_1 l_{BS2} \tag{1.36}$$

$$k_3 = m_2 l_{BS2}^2 + I_{S2} + I_{S3} + I_{S4} + I_{S5} + I_{S6} \tag{1.37}$$

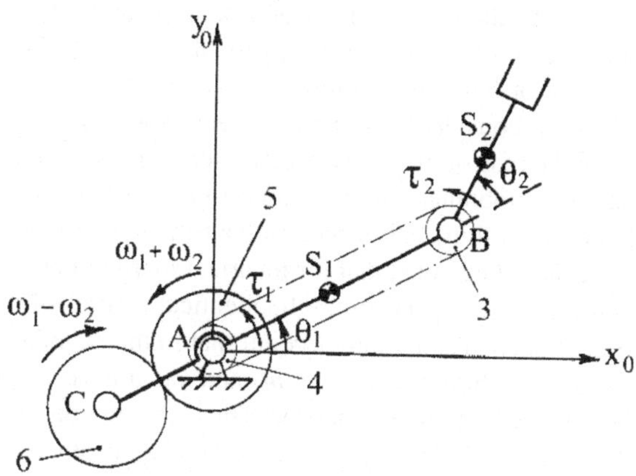

Fig. 1.6 Design of the decoupled 2-dof planar serial manipulator

$$k_4 = 2(k_3 - 2I_{S6}) \tag{1.38}$$

$$k_5 = (m_1 l_{AS1} + m_2 l_1 + m_3 l_1 - m_6 l_{AE})g \tag{1.39}$$

$$k_6 = m_2 l_{BS2} g \tag{1.40}$$

The static balancing of the manipulator leads to the condition:

$$k_2 = k_5 = k_6 = 0 \tag{1.41}$$

and

$$L = 0.5\left[(k_1)\dot{\theta}_1^2 + k_3\dot{\theta}_2^2 + (k_4)\dot{\theta}_1\dot{\theta}_2\right] \tag{1.42}$$

Thus, if $I_{S6} = 0.5k_3$, i.e. $k_4 = 0$, we obtain $L = 0.5\left(k_1\dot{\theta}_1^2 + k_3\dot{\theta}_2^2\right)$ and consequently

$$\tau_1 = k_1\ddot{\theta}_1 \tag{1.43}$$

and

$$\tau_2 = k_3\ddot{\theta}_2 \tag{1.44}$$

Now let us disclose the advantages with such a design: (*i*) gear 5 is mounted on the frame, which allows the reduction of the balancing masses; (*ii*) gear 6 is mounted on the prolongation of link 1, which allows the use of the mass of gear 6 designed for dynamic linearization as a counterweight for static balancing of link 1. Such a solution allows the considerable reduction of the total masses of links of the decoupled manipulator.

It is obvious that the proposed design can be used not only for the 2-DOF planar serial manipulators but also for any multi-degree-of-freedom spatial serial manipulators. In the case of the spatial manipulators, the suggested technique will take into account that the inertia tensor of the manipulator must be diagonal and independent of arm configuration.

Let us consider the dynamic linearization of a 3-DOF spatial serial manipulator. In the case of the 3-DOF spatial serial manipulator (Fig. 1.7) two rotating gears 7 and 8 are mounted on the second link of the manipulator and coupled with gear 6 linked with third link through a belt transmission.

It should be noted that for invariant inertia of any spatial serial manipulator there is the condition concerning the identical inertias of transverse direction, i.e. $I_x = I_y$ (see in [21]). Taking into account that in the examined spatial manipulator there are the links with masses m_2, m_3, m_7 and m_8, it is necessary that this condition will also be satisfied for these masses. Thus, the inertias of transverse direction will be identical, i.e. $I_m = I_{(m_i)x_2} = I_{(m_i)y_2}$. The kinetic energy of these masses relative to the

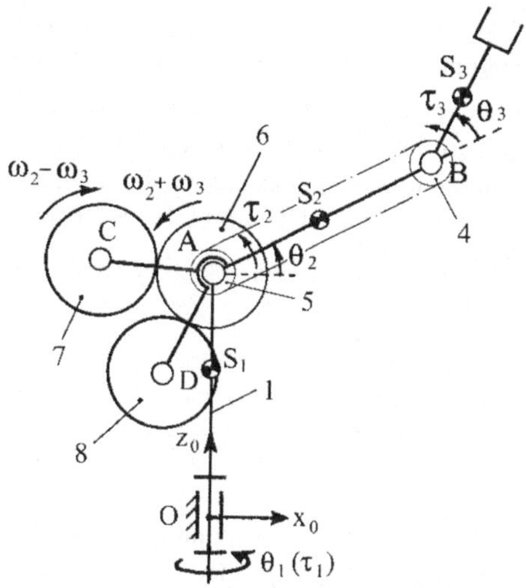

Fig. 1.7 Design of the decoupled 3-dof spatial serial manipulator

z_1 axis can be presented as $K_m = I_{(m_i)z_1}\dot{\theta}_1^2 = cI_m\dot{\theta}_1^2$, i.e. $I_{(m_i)z_1}$ can be replaced by the inertia cI_m, where c is a constant.

Thus, the Lagrangian of the statically balanced manipulator with the mentioned identical inertias of transverse direction of masses m_2, m_3, m_7 and m_8 $\left(I_m = I_{(m_i)x_2} = I_{(m_i)y_2}\right)$, can be written as:

$$\begin{aligned} L = 0.5\Big[& I_{z_1}\dot{\theta}_1^2 + cI_m\dot{\theta}_1^2 + I_m\dot{\theta}_2^2 + I_{x_2}\dot{\theta}_1^2\sin^2\theta_2 + I_{y_2}\dot{\theta}_1^2\cos^2\theta_2 \\ & + I_{z_2}\dot{\theta}_2^2 + I_{x_3}\dot{\theta}_1^2\sin^2(\theta_2+\theta_3) + I_{y_3}\dot{\theta}_1^2\cos^2(\theta_2+\theta_3) \\ & + I_{z_3}\left(\dot{\theta}_2+\dot{\theta}_3\right)^2 + (I_{z_7}+I_{z_8})\left(\dot{\theta}_2-\dot{\theta}_3\right)^2\Big] \\ = 0.5&\left[p_1\dot{\theta}_1^2 + p_2\dot{\theta}_2^2 + p_3\dot{\theta}_3^2 + p_4\dot{\theta}_1\dot{\theta}_2\right] \end{aligned} \tag{1.45}$$

with

$$p_1 = I_{z_1} + cI_m + I_{x_2}\sin^2\theta_2 + I_{y_2}\cos^2\theta_2 + I_{x_3}\sin^2(\theta_2+\theta_3) + I_{y_3}\cos^2(\theta_2+\theta_3) \tag{1.46}$$

$$p_2 = I_m + I_{z_2} + I_{z_3} + I_{z_7} + I_{z_8} \tag{1.47}$$

$$p_3 = I_{z_3} + I_{z_7} + I_{z_8} \tag{1.48}$$

$$p_4 = 2(I_{z_3} - I_{z_7} - I_{z_8}) \tag{1.49}$$

where I_{z_i} is the inertia of axial direction, I_{x_i} and I_{y_i} are the inertia of transverse directions with respect to the mass center of link i $(i = 1, 2, 3)$; θ_1 is the angular displacement of link 1 relative to the base; θ_2 is the angular displacement of link 2 relative to link 1; θ_3 is the angular displacement of link 3 relative to link 2; $\dot{\theta}_1$ is the angular velocity of link 1 relative to the base; $\dot{\theta}_2$ is the angular velocity of link 2 relative to link 1; $\dot{\theta}_3$ is the angular velocity of link 3 relative to link 2; c is a constant due to the distribution of masses m_2, m_3, m_7 and m_8 according to the mentioned condition of identical inertia about transverse directions.

From Eqs. (1.41) and (1.49) we can see that the manipulator dynamics will be decoupled if

$$I_{x_2} = I_{y_2} \tag{1.50}$$

$$I_{x_3} = I_{y_3} \tag{1.51}$$

$$I_{z_3} = I_{z_7} + I_{z_8} \tag{1.52}$$

By assuming these conditions, the kinetic energy of the manipulators will be constant and consequently

$$\tau_1 = p_1\ddot{\theta}_1 \tag{1.53}$$

$$\tau_2 = p_2\ddot{\theta}_2 \tag{1.54}$$

$$\tau_3 = p_3\ddot{\theta}_3 \tag{1.55}$$

Thus, the dynamic decoupling is again achieved.

In order to evaluate the dynamic performance of the improved design methodology let us consider a few illustrative examples.

For illustration of the performance of the suggested design methodology, the simulations using ADAMS software have been carried out for a 2-DOF and 3-DOF serial manipulators. The manipulator parameters of the 2-DOF serial manipulator (Fig. 1.6) are the following: $m_1 = 2.5$ kg; $m_2 = 2$ kg; $m_5 = 1$ kg; $m_6 = 6.5$ kg; $l_1 = 0.4$ m; $l_{AS1} = 0.2$ m; $l_{BS2} = 0$; $I_{S1} = 0.1$ kg m^2; $I_{S2} = 0.0275$ kg m^2; $I_{S3} = 0.005$ kg m^2; $I_{S5} = 0.0325$ kg m^2. The generation of motions between the initial and final positions of links: $\theta_{1I} = 0$; $\theta_{1F} = 0.5236$; $\theta_{2I} = 1.0472$ and $\theta_{2F} = 1.5708$, are carried out by the following fifth order polynomial laws: $\theta_j = 5.2360t^3 - 7.8540t^4 + 3.1416t^5$, $(j = 1, 2)$, $0 \le t \le 0.4$ s.

Figure 1.8 shows the variations of the input torques and accelerations. We can see that the dynamic equations are linearized, i.e. $\tau_1 = k_1\ddot{\theta}_1$ and $\tau_2 = k_3\ddot{\theta}_2$.

It should be also noted that the decoupled manipulator designed according to schematics proposed in [25] and having the same overall dimensions will be more than two times heavier.

Let us now consider the dynamic decoupling of the 3-DOF spatial serial manipulator (Fig. 1.7). The manipulator parameters are the following: $m_1 = 4$ kg; $m_2 = 2.5$ kg; $m_3 = 2$ kg; $m_6 = 1$ kg; $m_7 = 6.5$ kg; $m_8 = 6.5$ kg; $l_{AB} = l_2 = 0.4$ m; $l_{AS2} = 0.2$ m; $l_{BS3} = 0$; $l_{AC} = l_{AD} = 0.5612$ m; $\angle BAC = \angle BAD = 125°$;

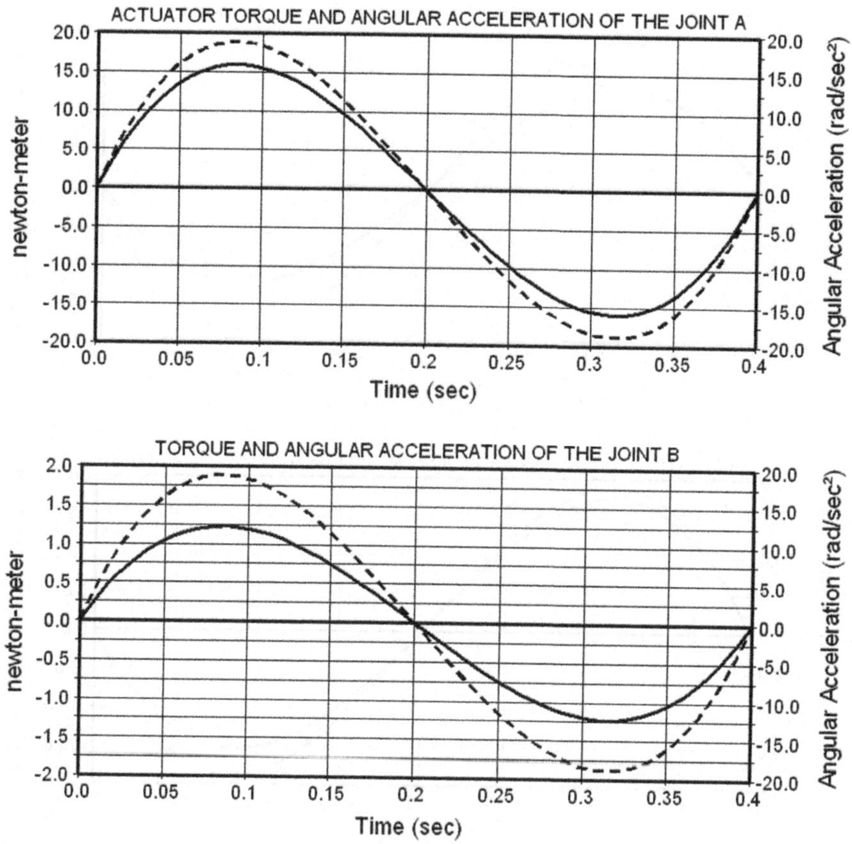

Fig. 1.8 Actuator torques (solid line) and angular accelerations (dashed line) for 2-dof linearized manipulator

$I_{z1} = 0.0125$ kg m^2; $I_{z2} = 0.1$ kg m^2; $I_{x2} = I_{y2} = 0.0175$ kg m^2; $I_{z3} = 0.0275$ kg m^2; $I_{x3} = I_{y3} = 0.01$ kg m^2; $I_{z5} = 0.005$ kg m^2; $I_{z7} = I_{z8} = 0.01625$ kg m^2. The generation of motions between the initial and final positions of links: $\theta_{1I} = 0$; $\theta_{1F} = 0.5236$; $\theta_{2I} = 1.0472$; $\theta_{2F} = 1.5708$, $\theta_{3I} = 0$; $\theta_{3F} = 0.5236$, are carried out by the following fifth order polynomial laws: $\theta_j = 5.2360t^3 - 7.8540t^4 + 3.1416t^5$, $(j = 1, 2, 3)$, $0 \leq t \leq 0.4$ s.

As shown in Fig. 1.9 the dynamic equations of the simulated manipulator are decoupled, i.e. $\tau_1 = p_1\ddot{\theta}_1$, $\tau_2 = p_2\ddot{\theta}_2$ and $\tau_3 = p_3\ddot{\theta}_3$.

Nevertheless, it should be noted that such a technique has a major disadvantage: the need for the connection of gears to the oscillating links. The gears added to the oscillating links of the manipulator are sources of shocks between teeth that will lead to the perturbation of the operation of the manipulator, and to noise and other negative effects.

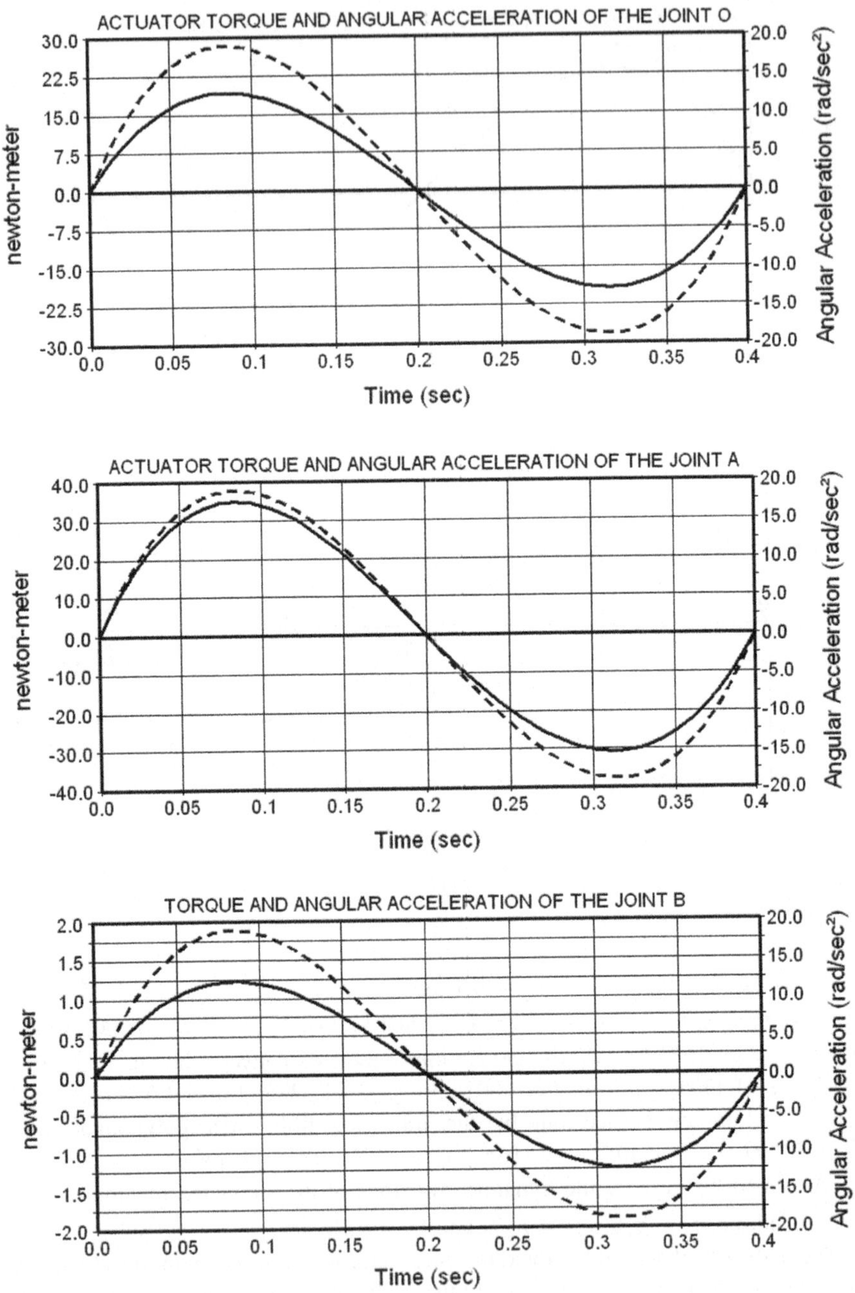

Fig. 1.9 Actuator torques (solid line) and angular accelerations (dashed line) for decoupled 3-dof serial manipulator

1.4 Dynamic Decoupling of Robot Manipulators by Using Epicyclic Gear Train

The proposed 2-DOF serial manipulator with parallel revolute joints differs from the traditional scheme by the fact that an epicyclic gear train is included between links 1 and 2 (Fig. 1.10).

In this design concept, gear $2'$ is connected with link 2 of the exoskeleton arm. The epicyclic gear train with sun gear $1'$ and epicyclic gears $3'$ and $3''$ is introduced between links 1 and 2 of the exoskeleton.

Let us rewrite the Lagrangian of the manipulator with the added links:

$$L = K - P = \sum_{i=1}^{4} K_i - \sum_{i=1}^{4} P_i = 0.5\left(m_1 V_{S1}^2 + m_2 V_{S2}^2 + m_3 V_{S3}^2 + m_4 V_{S4}^2\right)$$
$$+0.5\left[I_{S1}\dot{\theta}_1^2 + I_{S2}\left(\dot{\theta}_1 + \dot{\theta}_2\right)^2 + I_{S3}(\dot{\theta}_2 + \dot{\theta}_3)^2 + I_{S4}(\dot{\theta}_2 + \dot{\theta}_4)^2\right] - \sum_{i=1}^{4} P_i \quad (1.56)$$

where, θ_1 is the angular displacement of link 1 relative to the base; θ_2 is the angular displacement of link 2 relative to link 1; $\dot{\theta}_1$ is the angular velocity of link 1 relative to the base; $\dot{\theta}_2$ is the angular velocity of link 2 relative to link 1; m_1, m_2 are the masses of links 1 and 2; l_{AS_1} is the distance between the centre of mass S_1 of link 1 and joint centre A; l_{BS_2} is the distance between the centre of mass S_2 of link 2 and joint centre B, I_{S1} is the axial moment of inertia of link 1 relative to the centre of mass S_1 of link 1; I_{S2} is the axial moment of inertia of link 2 relative to the centre of mass S_2 of link 2; g is the gravitational acceleration; $\dot{\theta}_3$ and $\dot{\theta}_4$ are the angular velocities of links 3 and 4.

The static balancing of the exoskeleton arm leads to the condition $\sum_{i=1}^{4} P_i = 0$ and taking into consideration that the epicyclic gear train is symmetric regarding to the principal rotating axis, we obtain

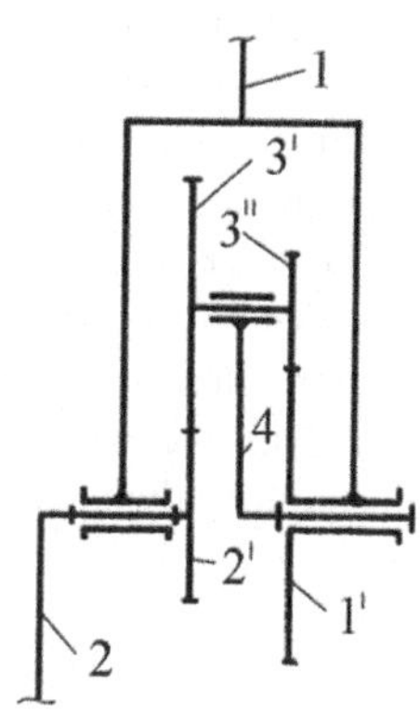

Fig. 1.10 The epicyclic gear train added between links 1 and 2

$$L = 0.5\left[(m_1 l_{AS1}^2 + I_{S1} + m l_1^2)\dot{\theta}_1^2 + I_{S2}\left(\dot{\theta}_1 + \dot{\theta}_2\right)^2 + I_{S3}(\dot{\theta}_1 + \dot{\theta}_3)^2 + I_{S4}(\dot{\theta}_1 + \dot{\theta}_4)^2\right] \quad (1.57)$$

where m is the mass of link 2 with epicyclic gear train.

Now, let us consider the transmission ratio in the epicyclic gear train:

$$\frac{\dot{\theta}_2}{\dot{\theta}_4} = 1 - \frac{Z_{3'}Z_{1'}}{Z_{2'}Z_{3''}} \quad (1.58)$$

where, $Z_{1'}, Z_{2'}, Z_{3'}$ and $Z_{3''}$ are the number of teeth of the corresponding gears.

Now, let us suggest that $Z_{3'} = 2Z_{2'}$ and $Z_{1'} = Z_{3'}$, which leads to the condition:

$$\dot{\theta}_2 = -\dot{\theta}_4 \quad (1.59)$$

With regard to the angular velocity of the planetary gear, we obtain

$$\dot{\theta}_3 = 2\dot{\theta}_4 = -2\dot{\theta}_2 \quad (1.60)$$

Introducing (1.59) and (1.60) into (1.57), we get

$$L = 0.5\left[(m_1 l_{AS1}^2 + I_{S1} + m l_1^2)\dot{\theta}_1^2 + I_{S2}\left(\dot{\theta}_1 + \dot{\theta}_2\right)^2 + I_{S3}(\dot{\theta}_1 - 2\dot{\theta}_2)^2 + I_{S4}(\dot{\theta}_1 - \dot{\theta}_2)^2\right] \quad (1.61)$$

The obtained equation can be rewritten as:

$$L = 0.5\left(k_1\dot{\theta}_1^2 + k_2\dot{\theta}_2^2 + k_3\dot{\theta}_1\dot{\theta}_2\right) \quad (1.62)$$

where, $k_1 = m_1 l_{AS1}^2 + m l_1^2 + I_{S1} + I_{S2} + I_{S3} + I_{S4}$; $k_2 = I_{S2} + 4I_{S3} + I_{S4}$ and $k_3 = 2(I_{S2} - 2I_{S3} - I_{S4})$.

Thus, if $k_3 = 0$, i.e. $I_{S2} = 2I_{S3} + I_{S4}$, we obtain $L = 0.5\left(k_1\dot{\theta}_1^2 + k_2\dot{\theta}_2^2\right)$ and consequently $\tau_1 = k_1\ddot{\theta}_1$ and $\tau_2 = k_2\ddot{\theta}_2$.

It should be noted that the epicyclic gear train provides very compact construction with high transmission ratio, which is a design advantage of the suggested decoupling technique.

In order to evaluate the performance of the suggested design concept let us consider an illustrative example.

For illustration of the suggested decoupling approach, simulations using ADAMS software have been carried. The exoskeleton arm parameters with decoupled dynamic equations are the following: $m_1 = 2$ kg, $m_2 = 8$ kg,

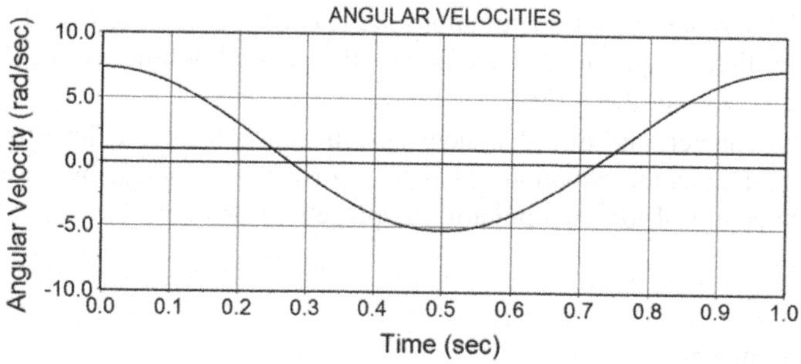

Fig. 1.11 Angular velocities of input links

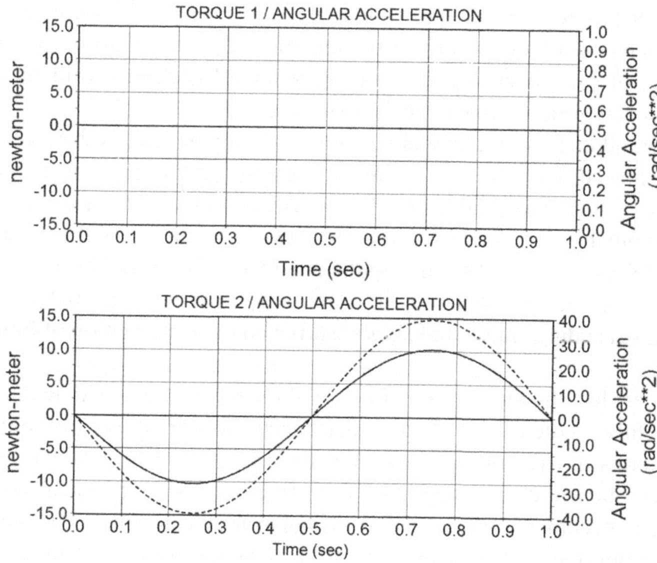

Fig. 1.12 Actuator torques (solid line) and angular accelerations (dashed line) for decoupled exoskeleton arm

$I_{S1} = I_{S2} = 0.1$ kg m^2, $l_{AS1} = 0.1$ m, $l_{BS2} = 0$ (taking into account that the epicyclic gear train is mainly a balanced system), $I_{S3} = 0.03$ kg m^2, $I_{S4} = 0.04$ kg m^2 (i.e. $I_{S2} = 2I_{S3} + I_{S4}$).

To better show the dynamic decoupling of motion equations the following input laws have been chosen: $\theta_1 = 60t$ and $\theta_2 = \sin(2\pi t)$ with $0 \le t \le 1$ s. Figure 1.11 shows the angular velocities of input links. Thus, the angular accelerations are $\ddot{\theta}_1 = 0$ and $\ddot{\theta}_2 = -4\pi^2 \sin(2\pi t)$.

Figure 1.12 shows the simulations results obtained by using ADAMS software, i.e. the variations of the input torques and accelerations. We can see that the dynamic decoupling of motion equations are carried, i.e. $\tau_1 = k_1\ddot{\theta}_1$ and $\tau_2 = k_3\ddot{\theta}_2$. Indeed, the first torque is cancelled and the second torque is the following: $\tau_2 = 0,26 \cdot [-4\pi^2 \sin(2\pi t)]$ (N).

This sub-chapter introduced a new design concept for the linearization and decoupling of dynamic equations of robot arms. It is achieved by adding to the initial architecture of the manipulator an epicyclic gear train.

1.5 Summary

In this chapter, a brief review of dynamic decoupling approaches has been presented. It should be noted that the dynamic decoupling via mass redistribution is simple and it found practical applications. However, as was mentioned above this solution can be used for dynamic decoupling of serial manipulators with orthogonal dispositions of joint axes. In the case of serial manipulators with parallel axes actuator relocation or auxiliary mechanisms are used.

However, all known mechanical solutions can only be reached by a considerably more complicated design of the initial structure via adding gears to the oscillating links leading to the unavoidable drawbacks. The gears added to the oscillating links of the manipulator are sources of shocks between teeth that will lead to the perturbation of the operation of the manipulator, and to noise and other negative effects.

Thus, it becomes evident that it is more optimal to carry out the dynamic decoupling via adding simple linkages ensuring an opposite motion of links without using gears.

The review has shown that the influence of the payload on the dynamic coupling remains lightly studied. It is obvious that mechanical solutions for adjustment of nonlinear terms of dynamic equations due to the changing payload can be reached by very complicated design solutions. It is not attractive for practical applications. Therefore, it becomes evident that it is indispensable to find new more simple solutions permitting to take into account the changing payload in the problems of dynamic decoupling.

On the other hand, dynamic decoupling via optimal control of a manipulator with a nonlinear system model and a changing payload is also rather complex task. That is why, it will be desirable to develop new approaches of dynamic decoupling, which are a symbiosis of mechanical and control solutions.

References

1. Descusse J, Lafay JF, Malabre M (1988) Solution to Morgan's problem. IEEE Trans Autom Control 33(8):732–739
2. Rocco P (1996) Stability of PID control for industrial robot arms. IEEE Trans Robot Autom 12(4):606–614

3. Santibanez V, Kelly R (1998) A class of nonlinear PID global regulators for robot manipulators. IEEE 3601–3606
4. Senthil Kumar J, Karthigai Amutha E (2014) Control and tracking of robotic manipulator using PID controller and hardware in Loop Simulation. IEEE 1–3
5. Su Y, Zheng C, Mullerz PC (2007) Global asymptotic stability of PID controller for robotic manipulators. 2007 international conference on mechatronics and automation 3244–3249
6. Craig JJ (2005) Introduction to robotics: mechanics and control. Pearson Prentice Hall Upper Saddle River
7. Ang KH, Chong G, Yun Li (2005) PID control system analysis, design, and technology. IEEE Trans Control Syst Technol 13(4):559–576
8. Ouyang PR, Pano V, Dam T (2015) PID position domain control for contour tracking. Int J Syst Sci 46(1):111–124
9. Asada H, Slotine J-JE (1986b) Robot analysis and control. Wiley
10. Lasalle JP Lefschetz (1961) Stability by Lyapunov's direct method. Academic Press New York
11. Ouyang PR, Zhang WJ, Gupta MM (2006) An adaptive switching learning control method for trajectory tracking of robot manipulators. Mechatronics 16(1):51–61
12. Atashzar SF, Talebi HA, Towhidkhah F (2010) A robust feedback linearization approach for tracking control of flexible-link manipulators using an EKF disturbance estimator. IEEE 1791–1796
13. Levine WS (1996) The control handbook. CRC Press
14. Hocking LM (1991) Optimal control: an introduction to the theory with applications. Clarendon Press
15. Yang HS, Slotine J-JE (1994) Fast algorithms for near-minimum-time control of robot manipulators communication. Int J Robot Res 13(6):521–532
16. Chen Y, Desrochers AA (1989) Structure of minimum-time control law for robotic manipulators with constrained paths. Proceedings of the IEEE international conference on robotics and automation 971–976
17. Rao VG, Bernstein DS (2001) Naive control of the double integrator. IEEE Control Syst 21 (5):86–97
18. Bhat SP, Bernstein DS, and others (1998) Continuous finite-time stabilization of the translational and rotational double integrators. IEEE Trans Autom Control 43(5):678–682
19. Slotine J-JE, Li W and others (1991) Applied nonlinear control. Prentice-Hall Englewood Cliffs, NJ
20. Arakelian V, Le Baron J-P, Mkrtchyan M (2016) Design of Scotch yoke mechanisms with improved driving dynamics. Proc Inst Mech Eng K: J Multi-Body Dyn 230(4):379–386
21. Youcef-Toumi K, Asada H (1986) The design of open-loop manipulator arms with decoupled and configuration-invariant inertia tensors. IEEE International Conference on Robotics and Automation 2018–2026
22. Youcef-Toumi K, Asada H (1987) The design of open-loop manipulator arms with decoupled and configuration-invariant inertia tensors. J Dyn Syst Meas Contr 109(3):268–275
23. Gompertz RS, Yang DC (1989) Performance evaluation of dynamically linearized and kinematically redundant planar manipulators. Robot comput-Integr Manuf 5(4):321–331
24. Belyanin PN, Konstantin F, Aron K, Alfred K (1981) Mechanical arm
25. Coelho TAH, Yong L, Alves VFA (2004) Decoupling of dynamic equations by means of adaptive balancing of 2-dof open-loop mechanisms. Mech Mach Theory 39(8):871–881

[illegible] (1986) [illegible]

Chapter 2
Design of Adjustable Serial Manipulators with Decoupled Dynamics

V. Arakelian, J. Xu and J. P. Le Baron

Abstract In this chapter, a new approach for dynamic decoupling of serial planar manipulators, which is a symbiosis of mechanical and control solutions is proposed. It is based on the opposite motion of manipulator links and the optimal command design. The opposite motion of links with optimal redistribution of masses allows the cancellation of the coefficients of nonlinear terms in the manipulator's kinetic and potential energy equations. Then, based on this completely linearized and decoupled manipulator, the simple linear control method is used. Furthermore, the changing payload is taken into account as a forward compensation in the controller. Finally, in order to stabilize the manipulator linearized and decoupled, a full state feedback is set up. The suggested design methodology is illustrated by simulations carried out using ADAMS and MATLAB software, which have confirmed the efficiency of the developed approach.

2.1 Design Concept of Manipulators with Adjustable Links

As mentioned, the inherent character of coupled dynamics is the prominent factor that impacts the operation accuracy and velocity performances of the serial manipulators. In this chapter, a mechatronic method is proposed to achieve the dynamic decoupling. The advantages of the suggested solution are:

V. Arakelian (✉) · J. Xu · J. P. Le Baron
Department of Mechanical and Control Systems Engineering, INSA,
20 Avenue des Buttes de Coësmes, CS 70839, 35708 Rennes Cedex, France
e-mail: vigen.arakelyan@insa-rennes.fr

V. Arakelian · J. P. Le Baron
LS2N/ECN, 1 rue de la Noë, BP 92101, 44321 Nantes Cedex 03, France

V. Arakelian (ed.), *Dynamic Decoupling of Robot Manipulators*, Mechanisms and Machine Science 56, https://doi.org/10.1007/978-3-319-74363-9_2

- a simple linkage is added for achieving the dynamic decoupling of the serial planar manipulators. It allows a dynamic decoupling of manipulators without connection of gears to the oscillating links of the manipulator having leading to imperfections.
- the simplification of the controller based on the linearized input/output relationships, hence the computational burden caused by the huge amount of iterative calculation is vanished. The real-time performance can be improved.
- the feasibility of the linear control method that used in this kind of manipulator.
- the changing payload can be taken into account based on the dynamic decoupling manipulator.

Figure 2.1 shows the proposed adjustable serial manipulator for ensuring the dynamic decoupling of motion equations.

It is composed of link 1 with elements 1a, 1b and link 2 with elements 2a, 2b. The adjustable links of the manipulator allow an optimal selection of the lengths $L_1 = L_{AB}$ and $L_2 = L_{BP}$ of links 1 and 2, which ensures an identical and opposite rotation of links.

It can also be seen that the proposed manipulator is provided with a double Scott–Russell mechanism, which ensures the static balancing of link 2 for any position of element 2b.

To disclose the operation of the proposed adjustable manipulators, let's first consider the dynamic decoupling of an arbitrary serial manipulator.

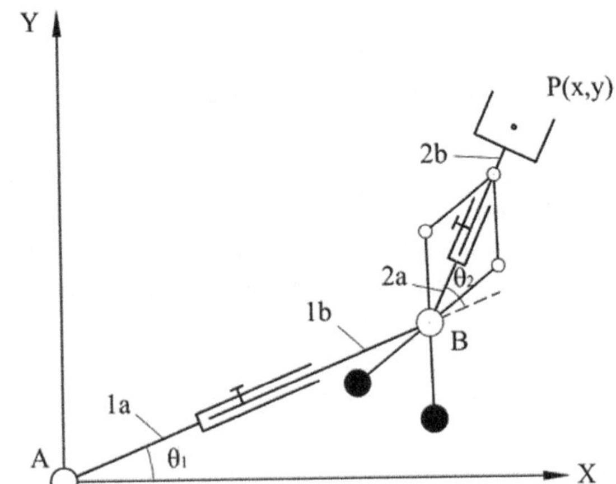

Fig. 2.1 Design concept of dynamically decoupled planar serial manipulator with adjustable links

2.1.1 Dynamic Decoupling Modeling of an Arbitrary Serial Manipulator

So let's start by the dynamic decoupling of a serial planar manipulator with two degrees of freedom shown in Fig. 2.2.

According to Lagrangian dynamics, the equations of motion can be written as

$$\begin{bmatrix}\tau_1\\ \tau_2\end{bmatrix}=\begin{bmatrix}D_{11} & D_{12}\\ D_{21} & D_{22}\end{bmatrix}\begin{bmatrix}\ddot{\theta}_1\\ \ddot{\theta}_2\end{bmatrix}+\begin{bmatrix}D_{111} & D_{122}\\ D_{211} & D_{222}\end{bmatrix}\begin{bmatrix}\dot{\theta}_1^2\\ \dot{\theta}_2^2\end{bmatrix}+\begin{bmatrix}D_{112} & D_{121}\\ D_{212} & D_{221}\end{bmatrix}\begin{bmatrix}\dot{\theta}_1\dot{\theta}_2\\ \dot{\theta}_1\dot{\theta}_2\end{bmatrix}+\begin{bmatrix}D_1\\ D_2\end{bmatrix} \tag{2.1}$$

with

$$D_{11}=m_1L_{AS1}^2+m_2L_1^2+m_2L_{BS2}^2+2m_2L_1L_{BS2}\cos\theta_2+I_{S1}+I_{S2} \tag{2.2}$$

$$D_{12}=D_{21}=m_2L_{BS2}^2+m_2L_1L_{BS2}\cos\theta_2+I_{S2} \tag{2.3}$$

$$D_{22}=m_2L_{BS2}^2+I_{S2} \tag{2.4}$$

$$D_{111}=0 \tag{2.5}$$

$$D_{122}=-m_2L_1L_{BS2}\sin\theta_2 \tag{2.6}$$

$$D_{211}=m_2L_1L_{BS2}\sin\theta_2 \tag{2.7}$$

$$D_{222}=0 \tag{2.8}$$

$$D_{112}=D_{121}=-m_2L_1L_{BS2}\sin\theta_2 \tag{2.9}$$

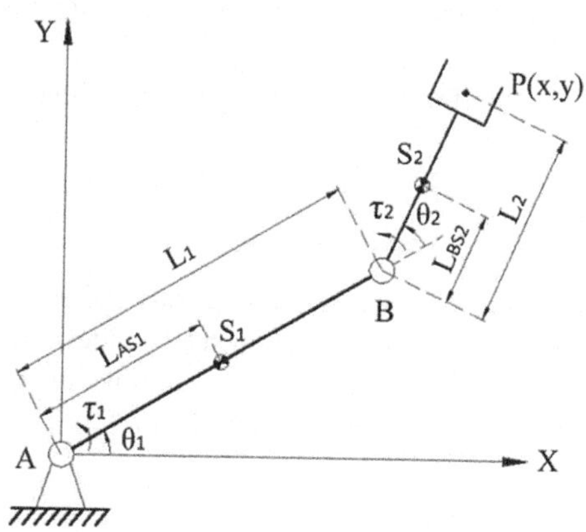

Fig. 2.2 An arbitrary planar serial manipulator

$$D_{212} = D_{221} = 0 \tag{2.10}$$

$$D_1 = (m_1 L_{AS1} + m_2 L_1) g \cos\theta_1 + m_2 g L_{BS2} \cos(\theta_1 + \theta_2) \tag{2.11}$$

$$D_2 = m_2 g L_{BS2} \cos(\theta_1 + \theta_2) \tag{2.12}$$

where τ_1 and τ_2 are respectively the actuator torques in A and B; L_1, L_2 are the lengths of links AB and BP; θ_1 is the angular displacement of link AB relative to the base; θ_2 is the angular displacement of link BP relative to link AB; $\dot{\theta}_1$ is the angular velocity of link AB relative to the base; $\dot{\theta}_2$ is the angular velocity of link BP relative to link AB; m_1, m_2 are the masses of links AB and BP; $\ddot{\theta}_1$ is the angular acceleration of link AB relative to the base; $\ddot{\theta}_2$ is the angular acceleration of link BP relative to link AB; m_1, m_2 are the masses of links AB and BP;L_{AS1} is the distance between the center of mass S_1 of link AB and joint center A; L_{BS2} is the distance between the center of mass S_2 of link BP and joint center B; I_{S1} is the axial moment of inertia of link AB relative to the center of mass S_1 of link AB; I_{S2} is the axial moment of inertia of link BP relative to the center of mass S_2 of link BP; g is the gravitational acceleration.

As known, this is a typical dynamic coupled and nonlinear model. In order to reduce or even eliminate the influence that cased by the coupling, the dynamic model must be improved further.

Now, let us consider that the second link is statically balanced, i.e. $L_{BS2} = 0$ and the gravitational forces are perpendicular to the motion plane xOy, i.e. $D_1 = D_2 = 0$.

With the conditions above, the Eq. (2.1) can be rewritten as

$$\begin{bmatrix} \tau_1 \\ \tau_2 \end{bmatrix} = \begin{bmatrix} a+b & a \\ a & a \end{bmatrix} \begin{bmatrix} \ddot{\theta}_1 \\ \ddot{\theta}_2 \end{bmatrix} \tag{2.13}$$

where, $a = I_{S2}$; $b = I_{S1} + m_1 L_{AS1}^2 + m_2 L_1^2$.

Obviously, this is linearized dynamic model but still uncoupled.

To ensure the dynamic decoupling, it is necessary to ensure the condition $\ddot{\theta}_1 = -\ddot{\theta}_2$.

Then the Eq. (2.13) can be simplified as

$$\begin{cases} \tau_1 = b\ddot{\theta}_1 \\ \tau_2 = 0 \end{cases} \tag{2.14}$$

Thus, we obtain a completely decoupled and linearized model.

Next step is the geometric synthesis of the mechanical structure which should ensure two identical motions of links in opposite directions for any given initial and final positions of the end-effector.

2.1.2 Adjustment Lengths of Links for Ensuring Opposite Rotations

According to the inverse kinematics of the planar serial manipulator shown in Fig. 2.1, the joint angles can be expressed as

$$\theta_1 = \tan^{-1}\left[\frac{y(L_1 + L_2\cos\theta_2) - xL_2\sin\theta_2}{x(L_1 + L_2\cos\theta_2) + yL_2\sin\theta_2}\right] \tag{2.15}$$

$$\theta_2 = \pm\cos^{-1}\left[\frac{x^2 + y^2 - L_1^2 - L_2^2}{2L_1L_2}\right] \tag{2.16}$$

where x and y are the coordinates of the end-effector in Fig. 2.2.

The given expressions show that for the same end-effector position there are two possible configurations of the manipulator called "elbow down" [configuration noted (1)] and "elbow up" [configuration noted (2)]. The fact that a manipulator has multiple solutions would be used for ensuring the dynamic decoupling. Two configurations of the manipulator corresponding to the initial end-effector position P_i and the final end-effector position P_f are shown in Fig. 2.3. As it has been mentioned above, the initial position of the end-effector can be found by the following solutions: $\theta^i_{1(1)}, \theta^i_{2(1)}$ "elbow down" solution, $\theta^i_{1(2)}, \theta^i_{2(2)}$ "elbow up" solution (not shown) and the final position of the end-effector by $\theta^f_{1(1)}, \theta^f_{2(1)}$ "elbow down" solution (not shown), $\theta^f_{1(2)}, \theta^f_{2(2)}$ "elbow up" solution. Thus, the links of the manipulator move in such a manner that in the initial end-effector position (P_i), where the configuration of the manipulator will correspond to the "elbow down"

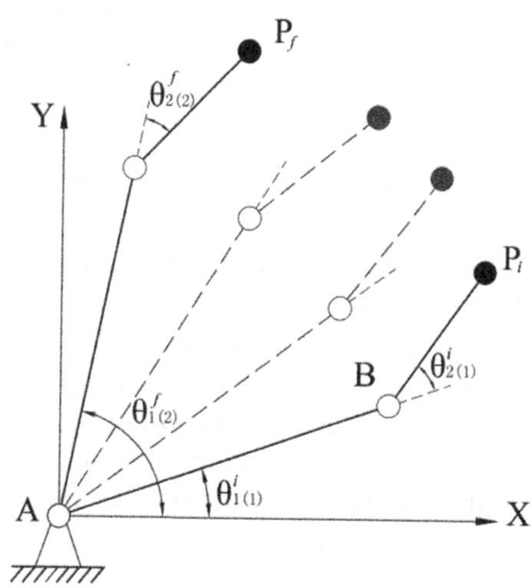

Fig. 2.3 Two configurations of the serial manipulator corresponding to the initial and final end-effector positions

solution; and, in the final end-effector position (P_f), where the configuration of the manipulator will correspond to the "elbow up" solution.

This choice of initial and final configurations of the manipulator with an optimal selection of lengths L_1 and L_2 allows equal ($\Delta\theta_1 = \Delta\theta_2$) and opposite ($\dot{\theta}_1 = -\dot{\theta}_2$) rotations of links AB and BP, i.e. $\left|\theta^f_{1(2)} - \theta^i_{1(1)}\right| = -\left|\theta^f_{2(2)} - \theta^i_{2(1)}\right|$. These conditions lead to $\ddot{\theta}_1 = -\ddot{\theta}_2$ and consequently to Eq. (2.14).

Now consider the selection of lengths L_1 and L_2 of links 1 and 2 for any given trajectory. To limit the variables in the specified conditions, suppose that the following parameters are given:

- the initial position P_i of the end-effector: x_i, y_i;
- the final position P_f of the end-effector: x_f, y_f;
- the initial angular position of the second link: $\theta^i_{2(1)}$;
- the rotating angle of the first link: $\Delta\theta_1 = \theta^f_{1(2)} - \theta^i_{1(1)}$.

The geometrical equations of the manipulator with the mentioned conditions lead to the following expressions:

$$L_1 = \frac{x_f^2 + y_f^2 - x_i^2 - y_i^2}{2L_2(\cos\theta^f_{2(2)} - \cos\theta^i_{2(1)})} \tag{2.17}$$

$$L_2 = \left[\frac{-\xi - (\xi^2 - 4\chi)^{1/2}}{2}\right]^{1/2} \tag{2.18}$$

where

$$\chi = \left[\frac{x_f^2 + y_f^2 - x_i^2 - y_i^2}{2(\cos\theta^f_{2(2)} - \cos\theta^i_{2(1)})}\right]^2 \tag{2.19}$$

$$\xi = 2(\chi)^{1/2}\cos\theta^i_{21} - x_i^2 - y_i^2 \tag{2.20}$$

$$\theta^f_{2(2)} = -(\Delta\theta_1 - \theta^i_{2(1)}) \tag{2.21}$$

Thus, for any initial and final positions of the end-effector, the lengths L_1 and L_2 determined from Eqs. (2.17) and (2.18) will ensure an equal ($\Delta\theta_1 = \Delta\theta_2$) and opposite ($\dot{\theta}_1 = -\dot{\theta}_2$) rotations of links AB and BP.

Let us now consider the Scott-Russell mechanism.

The Scott-Russell mechanism is usually used as a straight line generator, so it is also called exact straight line mechanism (Fig. 2.4). In this mechanism, the point C on the connecting rod copies the line traced by the slider B in a perpendicular direction OC, if OA = AB = AC. It should be noticed that the straight-line or linear

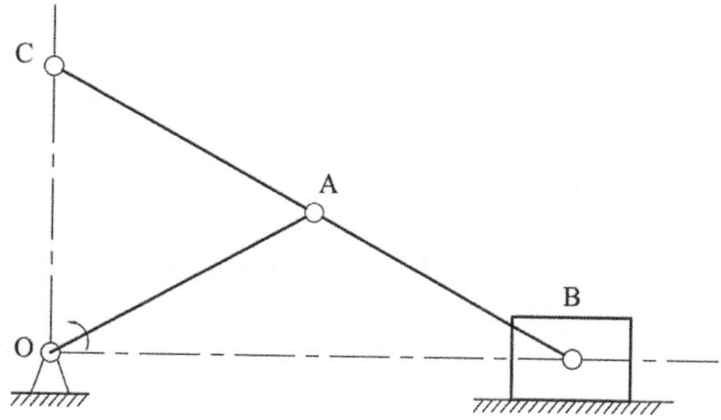

Fig. 2.4 Scott-Russell mechanism [1]

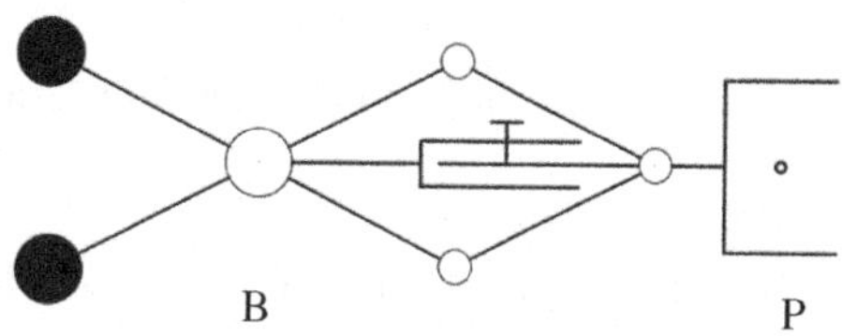

Fig. 2.5 Adjustable link BP with added Scott-Russell mechanisms

path exhibits great potential for high-speed pick and place operations in many manufacturing sectors [2].

As shown in Fig. 2.5, a pair of Scott-Russell mechanism is added. Thus, the proposed structure ensures the complete static balancing for any arbitrary configuration of the adjustable link BP.

2.2 Motion Generation and Dynamic Decoupling of the Adjustable Manipulators

In previous sections, we studied the kinematics and dynamics of the adjustable manipulator. This means that using the obtained equations of motion of the manipulator, we can determine the manipulator's positions.

The present section relates to the way the adjustable manipulator is moved from initial position to the final position, as well as the conditions of dynamic decoupling during the generated motions.

2.2.1 Motion Generation via Fifth-Order Polynomial Trajectory Planning

For motion generation between initial and final positions, a fifth-degree and secondary derivative polynomial is used

$$\theta = \sum_{i=0}^{5} c_i t^i \quad (i = 1, \ldots, 5) \tag{2.22}$$

Thus, we have

$$\theta(t) = c_0 + c_1 t + c_2 t^2 + c_3 t^3 + c_4 t^4 + c_5 t^5 \tag{2.23}$$

$$\dot{\theta}(t) = c_1 + 2c_2 t + 3c_3 t^2 + 4c_4 t^3 + 5c_5 t^4 \tag{2.24}$$

$$\ddot{\theta}(t) = 2c_2 + 6c_3 t + 12c_4 t^2 + 20c_5 t^3 \tag{2.25}$$

The initial and final conditions are given by $\begin{bmatrix} \theta(0) \\ \dot{\theta}(0) \\ \ddot{\theta}(0) \end{bmatrix} = \begin{bmatrix} \theta_I \\ 0 \\ 0 \end{bmatrix}$ and $\begin{bmatrix} \theta(T) \\ \dot{\theta}(T) \\ \ddot{\theta}(T) \end{bmatrix} = \begin{bmatrix} \theta_F \\ 0 \\ 0 \end{bmatrix}$, where T is the total time for the rotation. θ_I and θ_F are the initial and final positions in the joint space, respectively.

Substituting these conditions into Eqs. (2.23)–(2.25), we obtain

$$\theta(t) = \theta_I + [\theta_F - \theta_I]\left(\frac{t}{T}\right)^3 \left[10 - 15\left(\frac{t}{T}\right) + 6\left(\frac{t}{T}\right)^2\right] \tag{2.26}$$

$$\dot{\theta}(t) = [\theta_F - \theta_I]\frac{30}{T}\left(\frac{t}{T}\right)^2 \left[1 - 2\left(\frac{t}{T}\right) + \left(\frac{t}{T}\right)^2\right] \tag{2.27}$$

$$\ddot{\theta}(t) = [\theta_F - \theta_I]\frac{60}{T^2}\left(\frac{t}{T}\right) \left[1 - 3\left(\frac{t}{T}\right) + 2\left(\frac{t}{T}\right)^2\right] \tag{2.28}$$

To ensure the opposite rotation of links, the actuators generate the same motions. Hence, for joint A, we get

$$\begin{cases} \theta_1(t) = \theta_{1I} + [\theta_{1F} - \theta_{1I}]\left(\frac{t}{T}\right)^3 \left[10 - 15\left(\frac{t}{T}\right) + 6\left(\frac{t}{T}\right)^2\right] \\ \dot{\theta}_1(t) = [\theta_{1F} - \theta_{1I}]\frac{30}{T}\left(\frac{t}{T}\right)^2 \left[1 - 2\left(\frac{t}{T}\right) + \left(\frac{t}{T}\right)^2\right] \\ \ddot{\theta}_1(t) = [\theta_{1F} - \theta_{1I}]\frac{60}{T^2}\left(\frac{t}{T}\right) \left[1 - 3\left(\frac{t}{T}\right) + 2\left(\frac{t}{T}\right)^2\right] \end{cases} \tag{2.29}$$

for joint B, we get

$$\begin{cases} \theta_2(t) = \theta_{2I} + [\theta_{2F} - \theta_{2I}]\left(\frac{t}{T}\right)^3\left[10 - 15\left(\frac{t}{T}\right) + 6\left(\frac{t}{T}\right)^2\right] \\ \dot{\theta}_2(t) = [\theta_{2F} - \theta_{2I}]\frac{30}{T}\left(\frac{t}{T}\right)^2\left[1 - 2\left(\frac{t}{T}\right) + \left(\frac{t}{T}\right)^2\right] \\ \ddot{\theta}_2(t) = [\theta_{2F} - \theta_{2I}]\frac{60}{T^2}\left(\frac{t}{T}\right)\left[1 - 3\left(\frac{t}{T}\right) + 2\left(\frac{t}{T}\right)^2\right] \end{cases} \tag{2.30}$$

2.2.2 Dynamic Decoupling Without Payload

For the dynamic decoupling of the model, the Scott-Russell mechanism should be added. In consequence, the second link will be characterized by the link BP with the Scott-Russell mechanism and the counterweight.

The kinematic energies of all parts of the manipulator are:

$$E_1 = \frac{1}{2}m_1 L_{AS1}^2 \dot{\theta}_1^2 + \frac{1}{2}I_{S1}\dot{\theta}_1^2 \tag{2.31}$$

$$\begin{aligned} E_2 = \frac{1}{2}m_{2+SR+cw}[L_1^2\dot{\theta}_1^2 + L_{BS2r}^2(\dot{\theta}_1 + \dot{\theta}_2)^2 \\ + 2L_1 L_{BS2r}\cos(\theta_2)\dot{\theta}_1(\dot{\theta}_1 + \dot{\theta}_2)] + \frac{1}{2}I_{S2+SR+cw}(\dot{\theta}_1 + \dot{\theta}_2)^2 \end{aligned} \tag{2.32}$$

with,

$$\begin{aligned} m_{2+SR+cw} &= m_2 + m_{SR} + m_{cw} \\ L_{BS2r} &= \frac{m_2 L_{BS2} + m_{SR} L_{BS2} - m_{cw} L_{cw}}{m_{2+SR+cw}} \\ I_{S2+SR+cw} &= I_{S2} + I_{SR} + m_2(L_{BS2} - L_{BS2r})^2 \\ &\quad + m_{SR}(L_{BS2} - L_{BS2r})^2 + m_{cw}(L_{cw} + L_{BS2r})^2 \end{aligned} \tag{2.33}$$

where, m_2, m_{SR}, m_{cw} are the masses of link BP, Scott-Russell mechanism and counterweight, respectively; L_{BS2r} is referred to as the real distance between joint center B and center of mass of the object which is constituted by link BP, Scott-Russell mechanism and counterweight; $I_{S2+SR+cw}$ is referred to as the moment of inertia of this new combined object. In addition, the friction effect and gravity force are ignored here.

Thus, according to the Lagrangian equations, the torques of the manipulator without payload are:

$$\begin{aligned}\tau_{1without} &= \frac{d}{dt}\left(\frac{\partial L}{\partial \dot{\theta}_1}\right) - \frac{\partial L}{\partial \theta_1}\\ &= m_1 L_{AS1}^2 \ddot{\theta}_1 + I_{S1}\ddot{\theta}_1 + m_{2+SR+cw} L_1^2 \ddot{\theta}_1\\ &\quad + m_{2+SR+cw} L_{BS2r}^2 (\ddot{\theta}_1 + \ddot{\theta}_2) + m_{2+SR+cw} L_1 L_{BS2r} \cos(\theta_2)(2\ddot{\theta}_1 + \ddot{\theta}_2)\\ &\quad - m_{2+SR+cw} L_1 L_{BS2r} \sin(\theta_2)(2\dot{\theta}_1\dot{\theta}_2 + \dot{\theta}_2^2) + I_{S2+SR+cw}(\ddot{\theta}_1 + \ddot{\theta}_2)\end{aligned} \tag{2.34}$$

$$\begin{aligned}\tau_{2without} &= \frac{d}{dt}\left(\frac{\partial L}{\partial \dot{\theta}_2}\right) - \frac{\partial L}{\partial \theta_2}\\ &= m_{2+SR+cw} L_{BS2r}^2 (\ddot{\theta}_1 + \ddot{\theta}_2) + m_{2+SR+cw} L_1 L_{BS2r} \cos(\theta_2)\ddot{\theta}_1\\ &\quad + m_{2+SR+cw} L_1 L_{BS2r} \sin(\theta_2)\dot{\theta}_1^2 + I_{S2+SR+cw}(\ddot{\theta}_1 + \ddot{\theta}_2)\end{aligned} \tag{2.35}$$

where, $\tau_{1without}$ and $\tau_{2without}$ are torques of the manipulator without any payload.

Introducing the static equilibrium equations considered above, the dynamic equations can be simplified as

$$\begin{bmatrix}\tau_{1without}\\ \tau_{2without}\end{bmatrix} = \begin{bmatrix} a' + b' & a'\\ a' & a'\end{bmatrix}\begin{bmatrix}\ddot{\theta}_1\\ \ddot{\theta}_2\end{bmatrix} \tag{2.36}$$

where, $a' = I_{S2} + I_{SR} + m_2 L_{BS2}^2 + m_{SR} L_{BS2}^2 + m_{cw} L_{cw}^2$; $b' = m_1 L_{AS1}^2 + I_{S1} + m_{2+SR+cw} L_1^2$.

With the condition of opposite rotations ($\ddot{\theta}_2 = -\ddot{\theta}_1$), the Eq. (2.36) can be simplified as

$$\begin{cases}\tau_{1without} = b'\ddot{\theta}_1\\ \tau_{2without} = 0\end{cases} \tag{2.37}$$

2.3 Closed-Loop Control

As known, an open-loop system cannot correct any errors that it could make. And it will also not compensate for disturbances during the process. A closed-loop control system, also known as a feedback control system is a control system which uses the concept of an open-loop system as its forward path but has one or more feedback loops between its output and its input. A closed-loop control system is necessary when the open-loop system is unstable (double integrator).

Closed-loop systems are designed to automatically achieve and maintain the desired output trajectory by compensating an error signal which is the difference between the actual output and the reference input. Meanwhile, it can be used for compensating the disturbance during the while process.

According to the inverse calculation of the Eq. (2.36), the state equation of the MIMO model can be obtained with the state vector $\xi = [\theta_1 \quad \dot{\theta}_1 \quad \theta_2 \quad \dot{\theta}_2]^T$.

$$\underbrace{\begin{bmatrix} \dot{\xi}_1 \\ \dot{\xi}_2 \\ \dot{\xi}_3 \\ \dot{\xi}_4 \end{bmatrix}}_{\dot{\xi}} = \underbrace{\begin{bmatrix} 0 & 1 & 0 & 0 \\ 0 & 0 & 0 & 0 \\ 0 & 0 & 0 & 1 \\ 0 & 0 & 0 & 0 \end{bmatrix}}_{A} \underbrace{\begin{bmatrix} \xi_1 \\ \xi_2 \\ \xi_3 \\ \xi_4 \end{bmatrix}}_{\xi} + \underbrace{\begin{bmatrix} 0 & 0 \\ \frac{1}{b'} & -\frac{1}{b'} \\ 0 & 0 \\ -\frac{1}{b'} & \frac{1}{a'} + \frac{1}{b'} \end{bmatrix}}_{B} \underbrace{\begin{bmatrix} \tau_{1without} \\ \tau_{2without} \end{bmatrix}}_{\tau} \tag{2.38}$$

Before the further research, the controllable canonical form of the state equation should be built by the following transformation:

$$\xi = [T_{cc} W_c] \zeta \tag{2.39}$$

where, T_{cc} is the controllability matrix, and it is formed as

$$T_{cc} = [B|AB] = \begin{bmatrix} 0 & 0 & \frac{1}{b'} & -\frac{1}{b'} \\ \frac{1}{b'} & -\frac{1}{b'} & 0 & 0 \\ 0 & 0 & -\frac{1}{b'} & \frac{1}{a'} + \frac{1}{b'} \\ -\frac{1}{b'} & \frac{1}{a'} + \frac{1}{b'} & 0 & 0 \end{bmatrix}$$

W_c, is the upper triangular Toeplitz matrix, and it is formed by characteristic equation of matrix A: $|sI - A| = s^4 + a_3 s^3 + a_2 s^2 + a_1 s + a_0 = s^4$. So,

$$W_c = \begin{bmatrix} 1 & a_3 & a_2 & a_1 \\ 0 & 1 & a_3 & a_2 \\ 0 & 0 & 1 & a_3 \\ 0 & 0 & 0 & 1 \end{bmatrix} = \begin{bmatrix} 1 & 0 & 0 & 0 \\ 0 & 1 & 0 & 0 \\ 0 & 0 & 1 & 0 \\ 0 & 0 & 0 & 1 \end{bmatrix} = I$$

As known,

$$\dot{\xi} = A\xi + B\tau_{without} \tag{2.40}$$

Substituting the Eq. (2.39) into Eq. (2.40), we obtain

$$\dot{\zeta} = [T_{cc}]^{-1} A [T_{cc}] \zeta + [T_{cc}]^{-1} B \tau_{without} \tag{2.41}$$

Hence, the controllable canonical form is given by

$$\begin{bmatrix} \dot{\zeta}_1 \\ \dot{\zeta}_2 \\ \dot{\zeta}_3 \\ \dot{\zeta}_4 \end{bmatrix} = \begin{bmatrix} 0 & 0 & 0 & 0 \\ 0 & 0 & 0 & 0 \\ 1 & 0 & 0 & 0 \\ 0 & 1 & 0 & 0 \end{bmatrix} \begin{bmatrix} \zeta_1 \\ \zeta_2 \\ \zeta_3 \\ \zeta_4 \end{bmatrix} + \begin{bmatrix} 1 & 0 \\ 0 & 1 \\ 0 & 0 \\ 0 & 0 \end{bmatrix} \begin{bmatrix} \tau_{1without} \\ \tau_{2without} \end{bmatrix} \tag{2.42}$$

and the sate vector

$$\begin{bmatrix} \zeta_1 \\ \zeta_2 \\ \zeta_3 \\ \zeta_4 \end{bmatrix} = [T_{cc}]^{-1}\xi = \begin{bmatrix} (a' + b')\dot{\theta}_1 + a'\dot{\theta}_2 \\ a'\dot{\theta}_1 + a'\dot{\theta}_2 \\ (a' + b')\theta_1 + a'\theta_2 \\ a'\theta_1 + a'\theta_2 \end{bmatrix} \tag{2.43}$$

It is obviously that, the system that described by state Eq. (2.42) can be divided into two independent subsystems:

$$\begin{cases} \begin{bmatrix} \dot{\zeta}_1 \\ \zeta_3 \end{bmatrix} = \begin{bmatrix} 0 & 0 \\ 1 & 0 \end{bmatrix} \begin{bmatrix} \zeta_1 \\ \zeta_3 \end{bmatrix} + \begin{bmatrix} 1 \\ 0 \end{bmatrix} \tau_{1without} \\ \begin{bmatrix} \dot{\zeta}_2 \\ \zeta_4 \end{bmatrix} = \begin{bmatrix} 0 & 0 \\ 1 & 0 \end{bmatrix} \begin{bmatrix} \zeta_2 \\ \zeta_4 \end{bmatrix} + \begin{bmatrix} 1 \\ 0 \end{bmatrix} \tau_{2without} \end{cases} \tag{2.44}$$

Hence, when taking into account the feedback control, the closed-loop can be added firstly on each simple subsystem, then combine them together for the whole system. Obviously, both of these two subsystems are simple double integrator structures. Hence, the linear control method can be employed.

2.3.1 *Command of the First Double Integrator*

According to the first double integrator, we can deduce:

$$\begin{cases} \dot{\zeta}_1 = \tau_{1without} \\ \zeta_3 = \zeta_1 \end{cases} \tag{2.45}$$

Assuming that $\begin{cases} \rho = \zeta_3 \\ \dot{\rho} = \zeta_1 \end{cases}$

Then the state equation of the double integrator is given by

$$\begin{bmatrix} \dot{\rho} \\ \ddot{\rho} \end{bmatrix} = \underbrace{\begin{bmatrix} 0 & 1 \\ 0 & 0 \end{bmatrix}}_{A_1} \begin{bmatrix} \rho \\ \dot{\rho} \end{bmatrix} + \underbrace{\begin{bmatrix} 0 \\ 1 \end{bmatrix}}_{B_1} \tau_{1without} \tag{2.46}$$

Obviously, this is a simple double integrator structure. For the closed-loop control, the control law can be written as

$$\tau_{1without} = \mu_1 - g_{11}\dot{\rho} - g_{12}\rho \tag{2.47}$$

Substituting this control law in the Eq. (2.46), then the state equation of the double integrator can be rewritten as

$$\begin{bmatrix} \dot{\rho} \\ \ddot{\rho} \end{bmatrix} = \begin{bmatrix} 0 & 1 \\ -g_{12} & -g_{11} \end{bmatrix} \begin{bmatrix} \rho \\ \dot{\rho} \end{bmatrix} + \begin{bmatrix} 0 \\ 1 \end{bmatrix} \mu_1 \tag{2.48}$$

Assuming that

$$\mu_1 = \omega_1 + g_{11}\dot{\rho} + g_{12}\rho \tag{2.49}$$

then

$$\begin{bmatrix} \dot{\rho} \\ \ddot{\rho} \end{bmatrix} = \begin{bmatrix} 0 & 1 \\ 0 & 0 \end{bmatrix} \begin{bmatrix} \rho \\ \dot{\rho} \end{bmatrix} + \begin{bmatrix} 0 \\ 1 \end{bmatrix} \omega_1 \tag{2.50}$$

hence, $\omega_1 = \ddot{\rho}$. It can be found that

$$\mu_1 = \ddot{\rho} + g_{11}\dot{\rho} + g_{12}\rho \tag{2.51}$$

It is concluded that a balance position of the double integrator is given by the conditions about ρ and its continual derivatives. Hence, if a trajectory ρ_R which is secondary derivable over a range [0, T] is given, we can deduce:

$$\mu_{1R} = \ddot{\rho}_R + g_{11}\dot{\rho}_R + g_{12}\rho_R \tag{2.52}$$

Finally, the control law of the closed-loop double integrator is

$$\tau_{1without} = \ddot{\rho}_R - g_{11}[\dot{\rho} - \dot{\rho}_R] - g_{12}[\rho - \rho_R] \tag{2.53}$$

As known, $\rho = (a' + b')\theta_1 + a'\theta_2$, the control law can be rewritten as

$$\begin{aligned} \tau_{1without} = {} & (a' + b')\ddot{\theta}_{1R} + a'\ddot{\theta}_{2R} \\ & - g_{11}[(a' + b')[\dot{\theta}_1 - \dot{\theta}_{1R}] + a'[\dot{\theta}_2 - \dot{\theta}_{2R}]] \\ & - g_{12}[(a' + b')[\theta_1 - \theta_{1R}] + a'[\theta_2 - \theta_{2R}]] \end{aligned} \tag{2.54}$$

2.3.2 Feedback Parameters of the First Double Integrator

According to the last section, the double integration state equation of the first double integrator is obtained.

For realize the control law: $\tau_{1without} = -G_1[\rho \;\; \dot{\rho}]^T$, the matrix of state feedback parameter $G_1 = [g_{12} \;\; g_{11}]$ should be obtained. It is calculated by minimizing the index of performance J below:

$$J = \int_0^\infty \left[[\rho \;\; \dot{\rho}] Q_1 \begin{bmatrix} \rho \\ \dot{\rho} \end{bmatrix} + \tau_{1without}^T R_1 \tau_{1without} \right] dt \tag{2.55}$$

where

$$R_1 = 1 > 0$$

$$Q_1 = \left[T_{p1} \left(\int_0^{T_{p1}} e^{A_1 \tau} B_1 B_1^T e^{A_1^T \tau} d\tau \right) \right]^{-1} = \begin{bmatrix} \frac{12}{T_{p1}^4} & -\frac{6}{T_{p1}^3} \\ -\frac{6}{T_{p1}^3} & \frac{4}{T_{p1}^2} \end{bmatrix} > 0$$

Assuming that the symmetric matrix Σ_1 is the solution of Riccati equation

$$A_1^T \Sigma_1 + \Sigma_1 A_1 - \Sigma_1 B_1 R_1^{-1} B_1^T \Sigma_1 + Q_1 = 0 \tag{2.56}$$

where $\mathbf{\Sigma}_1 = \begin{bmatrix} \sigma_1 & \sigma_2 \\ \sigma_2 & \sigma_3 \end{bmatrix}$.

It is known that, the matrix will be positive if: $\begin{cases} \sigma_1 > 0 \\ \sigma_1 \sigma_3 - \sigma_2^2 > 0 \end{cases}$ thus

$$\sigma_1 = \sigma_2 \sigma_3 + \frac{6}{T_{p1}^3} > 0;$$

$$\sigma_2^2 = \frac{12}{T_{p1}^4} \Rightarrow \sigma_2 = \pm \frac{2\sqrt{3}}{T_{p1}^2};$$

$$\sigma_3^2 = 2\sigma_2 + \frac{4}{T_{p1}^2} > 0.$$

we obtain

$$\begin{cases} 2\sigma_2 + \frac{4}{T_{p1}^2} > 0 \Rightarrow \sigma_2 = \frac{2\sqrt{3}}{T_{p1}^2} \Rightarrow \sigma_3^2 = \frac{4(1+\sqrt{3})}{T_{p1}^2} \\ \sigma_2 \sigma_3 + \frac{6}{T_{p1}^3} > 0 \Rightarrow \sigma_3 = \frac{2\sqrt{1+\sqrt{3}}}{T_{p1}} \\ \sigma_1 = \sigma_2 \sigma_3 + \frac{6}{T_{p1}^3} = \frac{6+4\sqrt{3}\sqrt{1+\sqrt{3}}}{T_{p1}^3} \end{cases} \tag{2.57}$$

Hence, the expression of matrix Σ_1 is obtained

$$\Sigma_1 = \begin{bmatrix} \frac{6+4\sqrt{3}\sqrt{1+\sqrt{3}}}{T_{p1}^3} & \frac{2\sqrt{3}}{T_{p1}^2} \\ \frac{2\sqrt{3}}{T_{p1}^2} & \frac{2\sqrt{1+\sqrt{3}}}{T_{p1}} \end{bmatrix} \tag{2.58}$$

As known, the matrix G_1 is noted as

$$G_1 = R_1^{-1} B_1^T \Sigma_1 \tag{2.59}$$

Then,

$$G_1 = \begin{bmatrix} \frac{2\sqrt{3}}{T_{p1}^2} & \frac{2\sqrt{1+\sqrt{3}}}{T_{p1}} \end{bmatrix} \tag{2.60}$$

As result, we obtain $g_{12} = \frac{2\sqrt{3}}{T_{p1}^2}$ and $g_{11} = \frac{2\sqrt{1+\sqrt{3}}}{T_{p1}}$.

The state model of the closed-loop system is written as follows

$$\begin{bmatrix} \dot{\rho} \\ \ddot{\rho} \end{bmatrix} = [A_1 - B_1 G_1] \begin{bmatrix} \rho \\ \dot{\rho} \end{bmatrix} = \begin{bmatrix} 0 & 1 \\ -\frac{2\sqrt{3}}{T_{p1}^2} & -\frac{2\sqrt{1+\sqrt{3}}}{T_{p1}} \end{bmatrix} \begin{bmatrix} \rho \\ \dot{\rho} \end{bmatrix} \tag{2.61}$$

The characteristic polynomial P(s) is given by

$$P(s) = |sI - A_1 + B_1 G_1| = s^2 + \frac{2\sqrt{1+\sqrt{3}}}{T_{p1}} s + \frac{2\sqrt{3}}{T_{p1}^2} \tag{2.62}$$

If $P(s) = s^2 + 2\zeta\omega_n s + \omega_n^2$, we obtain $\omega_n = \frac{\sqrt{2\sqrt{3}}}{T_{p1}}$ and $\zeta = \frac{\sqrt{1+\sqrt{3}}}{\sqrt{2\sqrt{3}}} = 0.89$.

As shown in Eq. (2.44), the two subsystems have the same state equations. Hence, the control law and feedback parameters of the first double integrator that obtained above are also suitable for the second double integrator:

$$\begin{aligned} \tau_{2without} = {} & a'\ddot{\theta}_{1R} + a'\ddot{\theta}_{2R} \\ & - g_{21}[a'[\dot{\theta}_1 - \dot{\theta}_{1R}] + a'[\dot{\theta}_2 - \dot{\theta}_{2R}]] \\ & - g_{22}[a'[\theta_1 - \theta_{1R}] + a'[\theta_2 - \theta_{2R}]] \end{aligned} \tag{2.63}$$

where, $g_{22} = \frac{2\sqrt{3}}{T_{p2}^2}$ and $g_{21} = \frac{2\sqrt{1+\sqrt{3}}}{T_{p2}}$.

2.4 Dynamic Decoupling Taking into Account the Payload

The introduction of the payload leads to new loads on the actuators which are also nonlinear. Here, consider the payload also as one part of link BP. The energies of both of the two links are described as

$$
\begin{aligned}
E_1 &= \frac{1}{2} m_1 L_{AS1}^2 \dot{\theta}_1^2 + \frac{1}{2} I_{S1} \dot{\theta}_1^2 \\
E_2 &= \frac{1}{2} M_2 [L_1^2 \dot{\theta}_1^2 + L_{BS2r}^2 (\dot{\theta}_1 + \dot{\theta}_2)^2 + 2 L_1 L_{BS2r} \cos(\theta_2) \dot{\theta}_1 (\dot{\theta}_1 + \dot{\theta}_2)] \\
&\quad + \frac{1}{2} I_{S2+SR+cw+p} (\dot{\theta}_1 + \dot{\theta}_2)^2
\end{aligned} \tag{2.64}
$$

where,

$$
M_2 = m_{2+SR+cw} + m_p; \quad L_{BS2r} = \frac{m_2 L_{BS2} + m_{SR} L_{BS2} - m_{cw} L_{cw} + m_p L_2}{M_2};
$$

$$
\begin{aligned}
I_{S2+SR+cw+P} &= I_{S2} + I_{SR} + m_2 (L_{BS2} - L_{BS2r})^2 + m_{SR} (L_{BS2} - L_{BS2r})^2 \\
&\quad + m_{cw} (L_{cw} + L_{BS2r})^2 + m_p (L_2 - L_{BS2r})^2.
\end{aligned}
$$

According to the Lagrangian, the dynamic equations are

$$
\begin{aligned}
\tau_{1with} &= \frac{d}{dt}\left(\frac{\partial L}{\partial \dot{\theta}_1}\right) - \frac{\partial L}{\partial \theta_1} \\
&= m_1 L_{AS1}^2 \ddot{\theta}_1 + I_{S1} \ddot{\theta}_1 + M_2 L_1^2 \ddot{\theta}_1 + M_2 L_{BS2r}^2 (\ddot{\theta}_1 + \ddot{\theta}_2) \\
&\quad + M_2 L_1 L_{BS2r} \cos(\theta_2)(2\ddot{\theta}_1 + \ddot{\theta}_2) - M_2 L_1 L_{BS2r} \sin(\theta_2)(2\dot{\theta}_1 \dot{\theta}_2 + \dot{\theta}_2^2) \\
&\quad + I_{S2+SR+cw+p} (\ddot{\theta}_1 + \ddot{\theta}_2)
\end{aligned} \tag{2.65}
$$

$$
\begin{aligned}
\tau_{2with} &= \frac{d}{dt}\left(\frac{\partial L}{\partial \dot{\theta}_2}\right) - \frac{\partial L}{\partial \theta_2} \\
&= M_2 L_{BS2r}^2 (\ddot{\theta}_1 + \ddot{\theta}_2) + M_2 L_1 L_{BS2r} \cos(\theta_2) \ddot{\theta}_1 \\
&\quad + M_2 L_1 L_{BS2r} \sin(\theta_2) \dot{\theta}_1^2 + I_{S2+SR+cw+p} (\ddot{\theta}_1 + \ddot{\theta}_2)
\end{aligned} \tag{2.66}
$$

where, τ_{1with} and τ_{2with} are the output torque values of the first actuator and the second actuator respectively when taking into account the payload on the end-effector of the serial manipulator.

Substituting the Eq. (2.33) into these dynamic equations, then they can be simplified as

$$\begin{bmatrix} \tau_{1with} \\ \tau_{2with} \end{bmatrix} = \begin{bmatrix} m_1 L_{AS1}^2 + I_{S1} + M_2 L_1^2 + I_{S2} + I_{SR} & I_{S2} + I_{SR} \\ I_{S2} + I_{SR} & I_{S2} + I_{SR} \end{bmatrix} \begin{bmatrix} \ddot{\theta}_1 \\ \ddot{\theta}_2 \end{bmatrix}$$
$$+ \begin{bmatrix} (m_2 + m_{SR}) L_{BS2}^2 + m_{cw} L_{cw}^2 & (m_2 + m_{SR}) L_{BS2}^2 + m_{cw} L_{cw}^2 \\ (m_2 + m_{SR}) L_{BS2}^2 + m_{cw} L_{cw}^2 & (m_2 + m_{SR}) L_{BS2}^2 + m_{cw} L_{cw}^2 \end{bmatrix} \begin{bmatrix} \ddot{\theta}_1 \\ \ddot{\theta}_2 \end{bmatrix}$$
$$+ \begin{bmatrix} m_p L_2^2 + 2 m_p L_1 L_2 \cos(\theta_2) & m_p L_2^2 + m_p L_1 L_2 \cos(\theta_2) \\ m_p L_2^2 + m_p L_1 L_2 \cos(\theta_2) & m_p L_2^2 \end{bmatrix} \begin{bmatrix} \ddot{\theta}_1 \\ \ddot{\theta}_2 \end{bmatrix}$$
$$+ \begin{bmatrix} -2 m_p L_1 L_2 \sin(\theta_2) \dot{\theta}_2 & -m_p L_1 L_2 \sin(\theta_2) \dot{\theta}_2 \\ m_p L_1 L_2 \sin(\theta_2) \dot{\theta}_1 & 0 \end{bmatrix} \begin{bmatrix} \dot{\theta}_1 \\ \dot{\theta}_2 \end{bmatrix} \quad (2.67)$$

Introducing the condition of opposite rotations ($\ddot{\theta}_2 = -\ddot{\theta}_1$), then

$$\begin{bmatrix} \tau_{1with} \\ \tau_{2with} \end{bmatrix} = \begin{bmatrix} m_1 L_{AS1}^2 + I_{S1} + m_{2+SR+cw} L_1^2 & 0 \\ 0 & 0 \end{bmatrix} \begin{bmatrix} \ddot{\theta}_1 \\ \ddot{\theta}_2 \end{bmatrix}$$
$$+ \begin{bmatrix} m_p L_1^2 + m_P L_1 L_2 \cos(\theta_2) & 0 \\ m_p L_1 L_2 \cos(\theta_2) & 0 \end{bmatrix} \begin{bmatrix} \ddot{\theta}_1 \\ \ddot{\theta}_2 \end{bmatrix}$$
$$+ \begin{bmatrix} -m_p L_1 L_2 \sin(\theta_2) \dot{\theta}_2 & 0 \\ m_p L_1 L_2 \sin(\theta_2) \dot{\theta}_1 & 0 \end{bmatrix} \begin{bmatrix} \dot{\theta}_1 \\ \dot{\theta}_2 \end{bmatrix} \quad (2.68)$$

Compare Eq. (2.68) with Eq. (2.37), the torques that caused by the introduction of the payload can be found

$$\begin{bmatrix} \Delta\tau_1 \\ \Delta\tau_2 \end{bmatrix} = \begin{bmatrix} m_p L_1^2 + m_p L_1 L_2 \cos(\theta_2) & 0 \\ m_p L_1 L_2 \cos(\theta_2) & 0 \end{bmatrix} \begin{bmatrix} \ddot{\theta}_1 \\ \ddot{\theta}_2 \end{bmatrix}$$
$$+ \begin{bmatrix} -m_p L_1 L_2 \sin(\theta_2) \dot{\theta}_2 & 0 \\ m_p L_1 L_2 \sin(\theta_2) \dot{\theta}_1 & 0 \end{bmatrix} \begin{bmatrix} \dot{\theta}_1 \\ \dot{\theta}_2 \end{bmatrix} \quad (2.69)$$

Hence, the dynamic equation with payload can be rewritten as

$$\begin{cases} \tau_{1with} = \tau_{1without} + \Delta\tau_1 \\ \tau_{2with} = \tau_{2without} + \Delta\tau_2 \end{cases} \quad (2.70)$$

This part is referred to as payload compensation. Extracting the payload compensation from the dynamic equation of the model taking into account the payload, then the model can be treated still as a dynamic decoupling model. This is another advantage that building a dynamic decoupling model taking into account the changing payload.

Under this mechatronic method, the dynamic model can be greatly simplified and meanwhile achieve the dynamic decoupling. In addition, the changing payload is also considered. As a result, no matter there is payload or not, the simple linear control law can always be used.

2.5 Illustrative Example with the SIMULINK Block of MATLAB

In this section, a simulation model will be built in the SIMULINK block of the MATLAB. The performance of the proposed technique is examined. First, the inverse transformation of the dynamic model with payload must be found. Rewritten the Eq. (2.64) as fellows

$$\begin{cases} \tau_1 = \psi(\theta_2)\ddot{\theta}_1 + [\gamma_2 + \beta(\theta_2)]\ddot{\theta}_2 - 2\alpha(\theta_2)\dot{\theta}_1\dot{\theta}_2 - \alpha(\theta_2)\dot{\theta}_2^2 \\ \tau_2 = [\gamma_2 + \beta(\theta_2)]\ddot{\theta}_1 + \gamma_2\ddot{\theta}_2 + \alpha(\theta_2)\dot{\theta}_1^2 \end{cases} \tag{2.71}$$

where

$$\begin{aligned} \alpha(\theta_2) &= M_2 L_1 L_{BS2r}\sin(\theta_2); \\ \beta(\theta_2) &= M_2 L_1 L_{BS2r}\cos(\theta_2); \\ \gamma_1 &= I_{S1} + m_1 L_{AS1}^2; \\ \gamma_2 &= I_{S2+SR+cw+p} + M_2 L_{BS2r}^2; \\ \psi(\theta_2) &= \gamma_1 + \gamma_2 + M_2 L_1^2 + 2\beta(\theta_2). \end{aligned}$$

The inverse dynamic equations of the system are:

$$\begin{aligned} \ddot{\theta}_1 = {} & \frac{\gamma_2}{\Delta(\theta_2)}\tau_1 - \frac{[\gamma_2 + \beta(\theta_2)]}{\Delta(\theta_2)}\tau_2 + \frac{\gamma_2\alpha(\theta_2)}{\Delta(\theta_2)}\dot{\theta}_1\dot{\theta}_2 \\ & + \frac{\alpha(\theta_2)[\gamma_2 + \beta(\theta_2)]}{\Delta(\theta_2)}\dot{\theta}_1^2 + \frac{\gamma_2\alpha(\theta_2)}{\Delta(\theta_2)}(\dot{\theta}_1 + \dot{\theta}_2)\dot{\theta}_2 \end{aligned} \tag{2.72}$$

$$\ddot{\theta}_2 = -\frac{[\gamma_2 + \beta(\theta_2)]}{\Delta(\theta_2)}\tau_1 + \frac{\psi(\theta_2)}{\Delta(\theta_2)}\tau_2 - \frac{\alpha(\theta_2)[\gamma_2 + \beta(\theta_2)]}{\Delta(\theta_2)}\dot{\theta}_1\dot{\theta}_2 - \frac{\alpha(\theta_2)\psi(\theta_2)}{\Delta(\theta_2)}\dot{\theta}_1^2 - \frac{\alpha(\theta_2)[\gamma_2 + \beta(\theta_2)]}{\Delta(\theta_2)}(\dot{\theta}_1 + \dot{\theta}_2)\dot{\theta}_2 \quad (2.73)$$

where $\Delta(\theta_2) = \gamma_1\gamma_2 + M_2L_1^2I_{S2} + [\alpha(\theta_2)]^2 > 0$.

2.5.1 Simulation Model of Open-Loop Control System

In order to demonstrate the influence of the payload compensation, an open-loop control system is built. In the simulation model, it mainly contains two blocks, a controller block and a manipulator block, shown as Fig. 2.6. The controller block is used to provide the desired trajectories and the payload compensation. The manipulator block is used to simulate the real planar serial manipulator.

The simulation parametric values are obtained by using the ADAMS software. It should note that, this software is good at dynamic simulation and the model that built in this software is the same with the model in real word. Thus, these parametric values in Table 2.1 can be considered as the real parametric values of a real manipulator.

As shown, they can be divided into three mainly categories: mass parameter, length parameter and moment of inertia parameter.

It should be noted that, for simplify the calculation, the parametric values of counterweight are given by

$$\begin{aligned} m_{cw} &= m_2 + m_{SR} \\ L_{cw} &= L_{BS2} \end{aligned} \quad (2.74)$$

Until now, a whole simulation model of the open-loop control system can be built. For the operation of the model, the desired trajectories that described by the

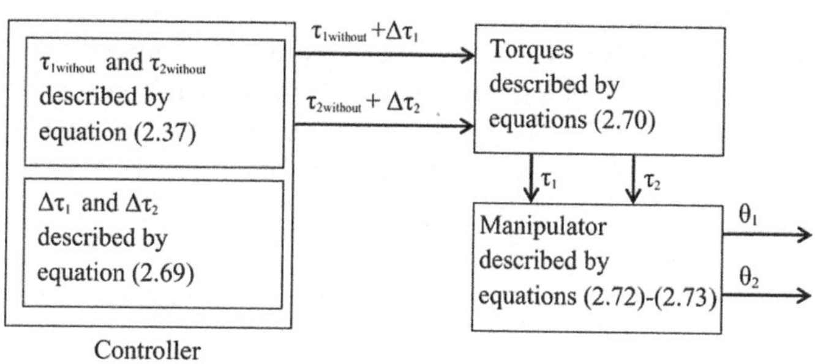

Fig. 2.6 The schema of the open-loop control system

Table 2.1 The parametric values of the open-loop control system

Mass (kg)		Length (m)		Moment of inertia (kg·m^2)	
m_1	13.193	L_1/L_{AS1}	0.8/0.4	I_{S1}	0.77754
m_2	8.477	L_2/L_{BS2}	0.5/0.25	I_{S2}	0.20744
m_{SR}	0.811	L_{BSR}	0.25	I_{SR}	0.01717
m_{cw}	9.288	L_{cw}	0.25		
m_p	5				

Table 2.2 The initial and final values of the desired trajectories

Angle (°)		Velocity (m/s)		Acceleration (m/s^2)	
θ_{1I}	33	$\dot{\theta}_{1I}$	0	$\ddot{\theta}_{1I}$	0
θ_{1F}	110	$\dot{\theta}_{1F}$	0	$\ddot{\theta}_{1F}$	0
θ_{2I}	0	$\dot{\theta}_{2I}$	0	$\ddot{\theta}_{2I}$	0
θ_{2F}	−77	$\dot{\theta}_{2F}$	0	$\ddot{\theta}_{2F}$	0

Eqs. (2.29) and (2.30) are used, and the initial and final positions of the manipulator are given as Table 2.2. Here, the total operation time $T = 1$ s.

According to the Eqs. (2.29) and (2.30), the desired trajectories, velocities and accelerations of the two actuators are shown in Fig. 2.7. Figure 2.7a1–a3 are the curves of the angular trajectory, angular velocity and angular acceleration of the first actuator respectively.

As the same, Fig. 2.7b1–b3 are the curves of the angular trajectory, angular velocity and angular acceleration of the second actuator respectively. It is obvious that these two trajectories are totally opposite. As proposed, this is a dynamic decoupling manipulator. The torque of each motor is only influence by one kind of parameter. Especially for the second actuator, its output torque is completely cancelled (Fig. 2.8).

In the controller of this system, the payload is considered as a forward compensation based on the linear and dynamic decoupling model. With payload compensation, both links of the manipulator can rotate exactly to the target angles, shown as the solid line in Fig. 2.9, because that the controller model reflects precisely the structure of the manipulator.

However, there are always some disturbances that can influence the precision. One of the extremely examples is cancellation of the payload compensation, shown as the dashed line in Fig. 2.9. Under this situation, the errors of angular displacements of link AB and BP are, respectively, 12.34 and 40.87%.

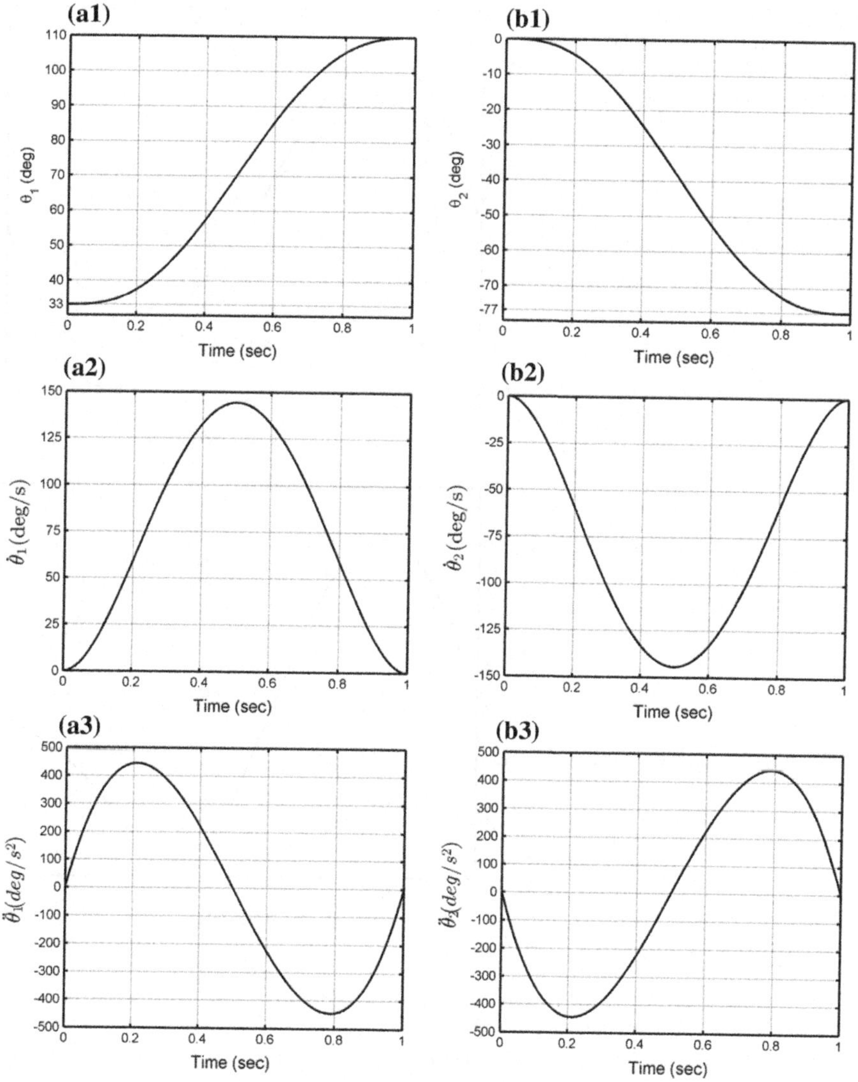

Fig. 2.7 Desired trajectories, velocities and accelerations of the two actuators

2.5.2 Simulation Model of Closed-Loop Control System

In order to reduce the impact of disturbance on tracking accuracy, the closed-loop is added. The closed-loop control law can be written as

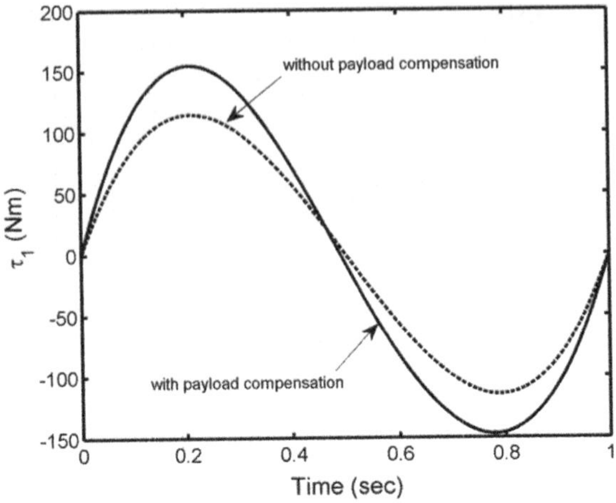

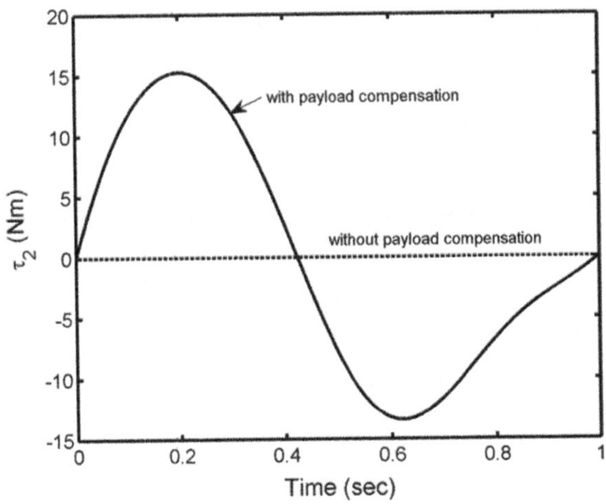

Fig. 2.8 Torques with payload compensation (solid line) and without it (dashed line) for the open-loop system

$$
\begin{aligned}
\tau_1 &= \tau_{1without} + \Delta\tau_1 \\
&\quad - g_{11}[(a' + b')[\dot{\theta}_1 - \dot{\theta}_{1R}] + a'[\dot{\theta}_2 - \dot{\theta}_{2R}]] \\
&\quad - g_{12}[(a' + b')[\theta_1 - \theta_{1R}] + a'[\theta_2 - \theta_{2R}]] \\
\tau_2 &= \tau_{2without} + \Delta\tau_2 \\
&\quad - g_{21}[a'[\dot{\theta}_1 - \dot{\theta}_{1R}] + a'[\dot{\theta}_2 - \dot{\theta}_{2R}]] \\
&\quad - g_{22}[a'[\theta_1 - \theta_{1R}] + a'[\theta_2 - \theta_{2R}]]
\end{aligned}
\tag{2.75}
$$

where, $\tau_{1without}$, $\tau_{2without}$, $\Delta\tau_1$ and $\Delta\tau_2$ are given by Eqs. (2.37) and (2.69). The simulation diagram in SIMULINK of MATLAB is shown in Fig. 2.10.

As in the case of open-loop system, the dashed curves show the torques and the angular displacements of the manipulator without payload compensation and the

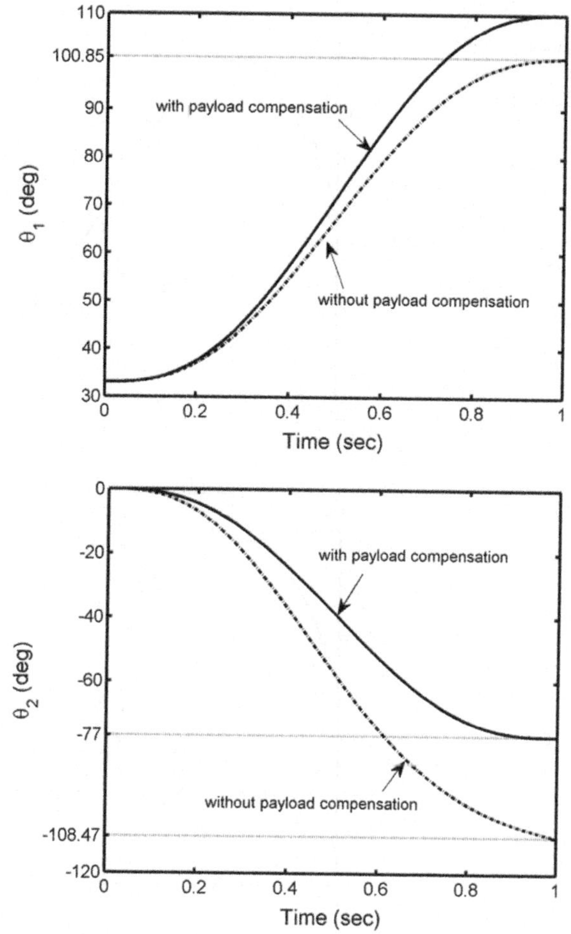

Fig. 2.9 Angular displacements of links with payload compensation (solid line) and without it (dashed line) for the open-loop system

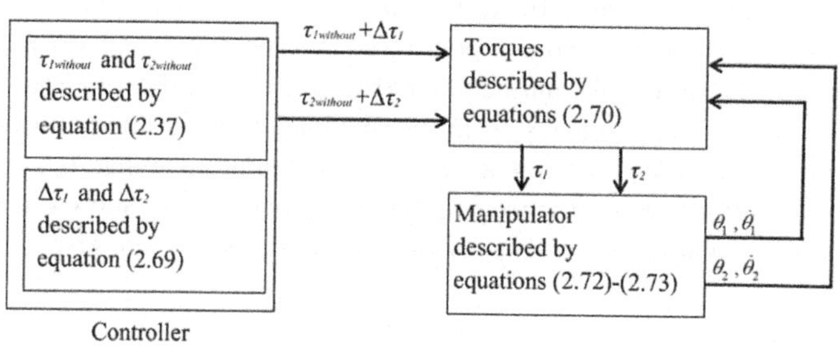

Fig. 2.10 The schema of the open-loop control system

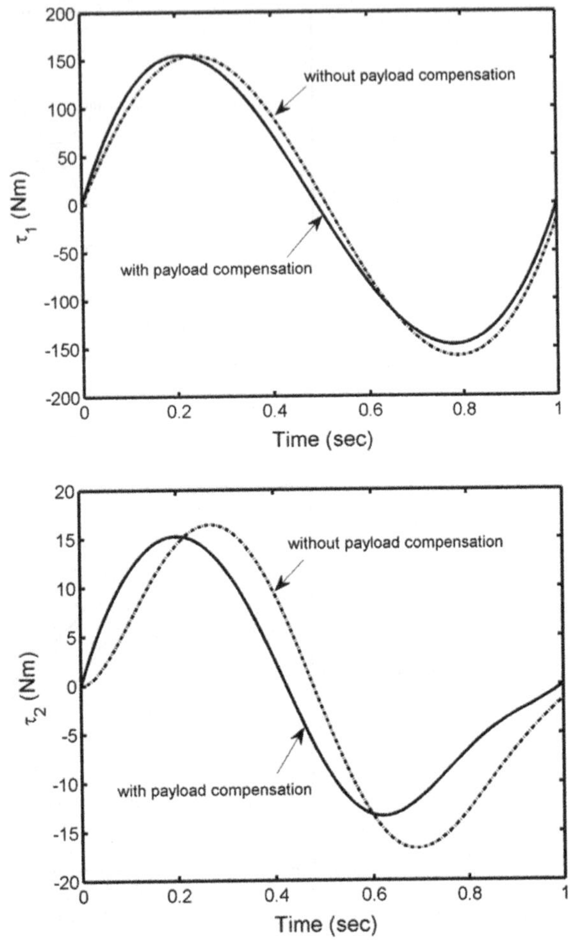

Fig. 2.11 Torques with payload compensation (solid line) and without it (dashed line) for the closed-loop system

solid curves show the simulation results of the model with payload compensation in Figs. 2.11 and 2.12, respectively.

Comparing the torque curves of closed-loop control system with the ones of open-loop control system, it is obviously that, the two torque curves of the simulation with and without payload compensation are closer in the closed-loop control system, especially for the second actuator (Fig. 2.11). In aspect of tracking accuracy, in this closed-loop control system, the payload compensation allows an exact reproduction of manipulator motions. However, it should note that, the closed-loop can reduce the influence of absence of payload compensation, shown in Fig. 2.12. The errors of angular displacements of link AB and link BP are respectively 1.43 and 1.18%.

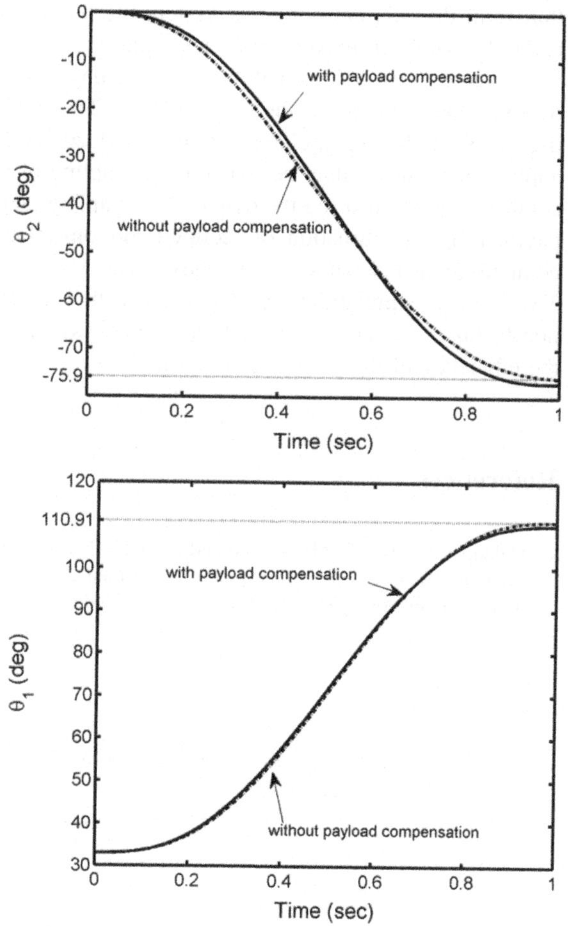

Fig. 2.12 Angular displacements of links with payload compensation (solid line) and without it (dashed line) for the closed-loop system

2.6 Summary

This chapter deals with the design concept of adjustable serial manipulators with linearized and decoupled dynamics taking into account the changing payload. It is achieved by using links with adjustable lengths connected to the double Scott-Russell mechanism and by means of an optimal control technique. Such a dynamic decoupling is a symbiosis of mechanical and control solutions. It is carried out in two steps. At first, the dynamic decoupling of the serial manipulator with adjustable lengths of links is accomplished via an opposite rotation of links and optimal redistribution of masses. Such a solution proposed for the first time allows one to carry out the dynamic decoupling without connection of gears to the oscillating links. The elimination of gears from design concept is a main advantage of the suggested solution. Thus, the proposed mechanical solution allows one to transform the original nonlinear system model into a fully linear system without using the feedback linearization technique.

However, it is obvious that the changing payload leads to the perturbation of the dynamic decoupling of the manipulator. To ensure linearized and decoupled dynamics of the manipulator for any payload, an optimal control technique is applied. It is shown that the dynamic decoupling of the manipulator simplifies the control solution ensuring the dynamic decoupling taking into account the changing payload. The perturbation of required motions of the manipulator with payload compensation and without it is shown via ADAMS and MATLAB simulations. Two kinds of simulations are carried out with open-loop control system which is a non-feedback system and closed-loop control system. The obtained results showed the efficiency of the proposed solution.

References

1. Dukkipati RV (2007) Mechanism and machine theory. Bohem press, Zürich
2. Liao G (2011) Design and analysis of a modified Scott Russell straight-line mechanism for a robot end-effector. J Appl Sci Eng Technol, 42–49

Chapter 3
Dynamic Decoupling of Planar Serial Manipulators with Revolute Joints

V. Arakelian, J. Xu and J. P. Le Baron

Abstract This chapter deals with a new dynamic decoupling principle, which involves connecting to a serial manipulator with revolute joints a two-link group forming a Scott-Russell mechanism with the initial links of the manipulator. The opposite motion of links in the Scott-Russell mechanism combined with optimal redistribution of masses allows the cancellation of the coefficients of nonlinear terms in the manipulator's kinetic and potential energy equations. Then, by using the optimal control design, the dynamic decoupling due to the changing payload is achieved. The suggested design methodology is illustrated by simulations carried out using ADAMS and MATLAB software, which have confirmed the efficiency of the developed approach.

3.1 Dynamic Decoupling Modeling via Adding a Two-Link Group

Figure 3.1 shows a serial planar manipulator with two degrees of freedom, which consists of two principal links 1, 2 and a sub-group with links 3 and slider 4. The movements of this manipulator are planar motions which are perpendicular to the vertical plane, and therefore, not subjected to gravitational forces. The slider 4 can slide freely along the link 1, and it's connected with link 3 by revolute joint D.

Thus, the added sub-group with link 2 forms a Scott-Russell mechanism [1]. As said above, the Scott-Russell mechanism has been developed to generate a rectilinear motion. Here, another property of this mechanism is used. The Scott-Russell linkage generates also rotations of links by identic angular accelerations, i.e. the angular accelerations of links 2 and 3 are similar.

V. Arakelian (✉) · J. Xu · J. P. Le Baron
Department of Mechanical and Control Systems Engineering, INSA,
20 avenue des Buttes de Coësmes, CS 70839, 35708 Rennes Cedex, France
e-mail: vigen.arakelyan@insa-rennes.fr

V. Arakelian · J. P. Le Baron
LS2N/ECN, 1 rue de la Noë, BP 92101, 44321 Nantes Cedex 03, France

V. Arakelian (ed.), *Dynamic Decoupling of Robot Manipulators*, Mechanisms and Machine Science 56, https://doi.org/10.1007/978-3-319-74363-9_3

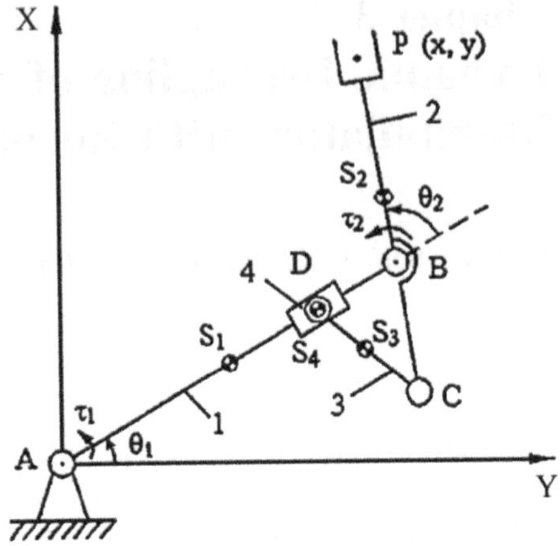

Fig. 3.1 The 2-dof planar serial manipulator with added two-link group

Let us consider the equations of motion of the unbalanced mechanism. In this case, the Lagrangian of the system is equal to the total kinetic energy. The kinetic energy expressions of all parts of the structure are:

$$E_1 = \frac{1}{2} m_1 L_{AS1}^2 \dot{\theta}_1^2 + \frac{1}{2} I_{S1} \dot{\theta}_1^2 \tag{3.1}$$

$$\begin{aligned} E_2 = {} & \frac{1}{2} m_2 [L_1^2 \dot{\theta}_1^2 + L_{BS2}^2 (\dot{\theta}_1 + \dot{\theta}_2)^2 + 2 L_1 L_{BS2} \cos(\theta_2) \dot{\theta}_1 (\dot{\theta}_1 + \dot{\theta}_2)] \\ & + \frac{1}{2} I_{S2} (\dot{\theta}_1 + \dot{\theta}_2)^2 \end{aligned} \tag{3.2}$$

$$\begin{aligned} E_3 = {} & \frac{1}{2} m_3 [L_1^2 \dot{\theta}_1^2 + L_3^2 (\dot{\theta}_1 + \dot{\theta}_2)^2 + L_{CS3}^2 (\dot{\theta}_1 - \dot{\theta}_2)^2 - 2 L_1 L_3 \cos(\theta_2) \dot{\theta}_1 (\dot{\theta}_1 + \dot{\theta}_2) \\ & - 2 L_1 L_{CS3} \cos(\theta_2) \dot{\theta}_1 (\dot{\theta}_1 - \dot{\theta}_2) + 2 L_3 L_{CS3} \cos(2\theta_2) (\dot{\theta}_1^2 - \dot{\theta}_2^2)] \\ & + \frac{1}{2} I_{S3} (\dot{\theta}_1 - \dot{\theta}_2)^2 \end{aligned} \tag{3.3}$$

$$E_4 = \frac{1}{2} m_4 [(2 L_3 \cos(\theta_2) - L_1)^2 \dot{\theta}_1^2 + 4 L_3^2 \sin^2(\theta_2) \dot{\theta}_2^2] + \frac{1}{2} I_{S4} \dot{\theta}_1^2 \tag{3.4}$$

Here, E_1, E_2, E_3 and E_4 are the kinetic energy of link 1, link 2, link 3 and slider 4 respectively. Then the dynamic equation of the serial planar manipulator can be obtained

$$\begin{bmatrix} \tau_1 \\ \tau_2 \end{bmatrix} = \begin{bmatrix} D_{11} & D_{12} \\ D_{21} & D_{22} \end{bmatrix} \begin{bmatrix} \ddot{\theta}_1 \\ \ddot{\theta}_2 \end{bmatrix} + \begin{bmatrix} D_{111} & D_{122} \\ D_{211} & D_{222} \end{bmatrix} \begin{bmatrix} \dot{\theta}_1^2 \\ \dot{\theta}_2^2 \end{bmatrix} + \begin{bmatrix} D_{112} & D_{121} \\ D_{212} & D_{221} \end{bmatrix} \begin{bmatrix} \dot{\theta}_1 \dot{\theta}_2 \\ \dot{\theta}_1 \dot{\theta}_2 \end{bmatrix} \tag{3.5}$$

with

$$D_{11} = m_1 L_{AS1}^2 + I_{S1} + m_2 L_1^2 + m_2 L_{BS2}^2 + 2m_2 L_1 L_{BS2}\cos\theta_2 + I_{S2} + m_3 L_1^2 + m_3 L_3^2 + m_3 L_{CS3}^2 - 2m_3 L_1 L_3 \cos\theta_2 - 2m_3 L_1 L_{CS3}\cos\theta_2 + 2m_3 L_3 L_{CS3}\cos(2\theta_2) + I_{S3} + m_4(2L_3\cos\theta_2 - L_1)^2 + I_{S4} \tag{3.6}$$

$$D_{12} = D_{21} = m_2 L_{BS2}^2 + m_2 L_1 L_{BS2}\cos\theta_2 + I_{S2} + m_3 L_3^2 - m_3 L_{CS3}^2 - m_3 L_1 L_3\cos\theta_2 + m_3 L_1 L_{CS3}\cos\theta_2 - I_{S3} \tag{3.7}$$

$$D_{22} = m_2 L_{BS2}^2 + I_{S2} + m_3 L_3^2 + m_3 L_{CS3}^2 - 2m_3 L_3 L_{CS3}\cos(2\theta_2) + I_{S3} + 4m_4 L_3^2 \sin^2\theta_2 \tag{3.8}$$

$$D_{111} = 0 \tag{3.9}$$

$$D_{122} = -m_2 L_1 L_{BS2}\sin\theta_2 + m_3 L_1 L_3 \sin\theta_2 - m_3 L_1 L_{CS3}\sin\theta_2 \tag{3.10}$$

$$D_{211} = m_2 L_1 L_{BS2}\sin\theta_2 - m_3 L_1 L_3\sin\theta_2 - m_3 L_1 L_{CS3}\sin\theta_2 + 2m_3 L_3 L_{CS3}\sin(2\theta_2) + 2m_4 L_3(2L_3\cos\theta_2 - L_1)\sin\theta_2 \tag{3.11}$$

$$D_{222} = 2m_3 L_3 L_{CS3}\sin(2\theta_2) + 2m_4 L_3^2\sin(2\theta_2) \tag{3.12}$$

$$D_{112} = -2m_2 L_1 L_{BS2}\sin\theta_2 + 2m_3 L_1 L_3\sin\theta_2 + 2m_3 L_1 L_{CS3}\sin\theta_2 - 4m_3 L_3 L_{CS3}\sin(2\theta_2) - 4m_4 L_3^2\sin(2\theta_2) + 4m_4 L_1 L_3 \sin(\theta_2) \tag{3.13}$$

$$D_{121} = D_{212} = D_{221} = 0 \tag{3.14}$$

where, m_1, m_2, m_3, and m_4 are the masses of link 1, link 2, link 3 and slider 4 respectively; I_{S1}, I_{S2}, I_{S3} and I_{S4} are the moments of inertia of link 1, link 2, link 3 and slider respectively; L_1, L_2, L_3 are the lengths of link 1, link 2, link 3 respectively, and L_3 also the distance between centers of revolute joints B and C; L_{AS1} is the distance between the center of mass S_1 of link 1 and joint center A; L_{BS2} is the distance between the center of mass S_2 of link 2 and joint center B; L_{CS3} is the distance between the center of joint C and center of mass of link 3; θ_1 is angular displacement of link 1 relative to the base; θ_2 is angular displacement of link 2 relative to link 1; $\dot{\theta}_1$ is angular velocity of link 1 relative to the base; $\dot{\theta}_2$ is angular velocity of link 2 relative to link 1; $\ddot{\theta}_1$ is the angular acceleration of link 1 relative to the base; $\ddot{\theta}_2$ is the angular acceleration of link 2 relative to link 1.

In order to further simplify this dynamic equation, two balancing conditions are applied [2]:

$$m_3 L_{CS3} + m_4 L_3 = 0 \tag{3.15}$$

$$(m_3 + m_4)L_3 - m_2 L_{BS2} = 0 \tag{3.16}$$

Substituting these two equations into the dynamic Eq. (3.5), then it can be simplified as

$$\begin{cases} \tau_1 = d\ddot{\theta}_1 + b\ddot{\theta}_2 \\ \tau_2 = b\ddot{\theta}_1 + c\ddot{\theta}_2 \end{cases} \tag{3.17}$$

where

$$\begin{aligned} d =& I_{S1} + I_{S2} + I_{S3} + I_{S4} + m_1 L_{AS1}^2 + (m_2 + m_3 + m_4) L_1^2 \\ &+ \left[\frac{(m_3 + m_4)^2}{m_2} + \frac{(m_3 + m_4)^2}{m_3} \right] L_3^2 \\ b =& I_{S2} - I_{S3} + \left[\frac{(m_3 + m_4)^2}{m_2} + \frac{m_3^2 - m_4^2}{m_3} \right] L_3^2 \\ c =& I_{S2} + I_{S3} + \left[\frac{(m_3 + m_4)^2}{m_2} + \frac{(m_3 + m_4)^2}{m_3} \right] L_3^2 \end{aligned}$$

Obviously, we obtain linear dynamic equations.

Furthermore, if

$$I_{S3} = I_{S2} + \left[\frac{(m_3 + m_4)^2}{m_2} + \frac{m_3^2 - m_4^2}{m_3} \right] L_3^2 \tag{3.18}$$

then we get $b = 0$.

Thus, the complete dynamic decoupling of the serial planar manipulator without payload is achieved:

$$\begin{cases} \tau_{1without} = d'\ddot{\theta}_1 \\ \tau_{2without} = c'\ddot{\theta}_2 \end{cases} \tag{3.19}$$

where,

$$\begin{aligned} d' =& I_{S1} + 2I_{S2} + I_{S4} + m_1 L_{AS1}^2 + (m_2 + m_3 + m_4) L_1^2 \\ &+ 2\frac{(m_3 + m_4)(m_2 + m_3 + m_4)}{m_2} L_3^2 \\ c' =& 2[I_{S2} + \frac{(m_3 + m_4)(m_2 + m_3 + m_4)}{m_2} L_3^2] \end{aligned}$$

3.2 Closed-Loop Control

The main goal of the closed-loop control is actually to stabilize this serial planar manipulator which is instable. The feedback design allows good tracking properties.

Define the state vector as $\xi = \begin{bmatrix} \theta_1 & \dot{\theta}_1 & \theta_2 & \dot{\theta}_2 \end{bmatrix}^{\mathrm{T}}$.

According to the inverse calculation of the Eq. (3.19), the state equation of the MIMO model can be obtained

$$\begin{bmatrix} \dot{\xi}_1 \\ \dot{\xi}_2 \\ \dot{\xi}_3 \\ \dot{\xi}_4 \end{bmatrix} = \underbrace{\begin{bmatrix} 0 & 1 & 0 & 0 \\ 0 & 0 & 0 & 0 \\ 0 & 0 & 0 & 1 \\ 0 & 0 & 0 & 0 \end{bmatrix}}_{A} \begin{bmatrix} \xi_1 \\ \xi_2 \\ \xi_3 \\ \xi_4 \end{bmatrix} + \underbrace{\begin{bmatrix} 0 & 0 \\ \frac{1}{d'} & 0 \\ 0 & 0 \\ 0 & \frac{1}{c'} \end{bmatrix}}_{B} \underbrace{\begin{bmatrix} \tau_{1without} \\ \tau_{2without} \end{bmatrix}}_{\tau} \tag{3.20}$$

The controllable canonical form of the state equation should be built for the further research. It needs to use the following transformation:

$$\xi = [T_{cc}\, W_c]\zeta \tag{3.21}$$

where, T_{cc} is the controllability matrix, and it is formed as

$$T_{cc} = [B|A \cdot B] = \begin{bmatrix} 0 & 0 & \frac{1}{d'} & 0 \\ \frac{1}{d'} & 0 & 0 & 0 \\ 0 & 0 & 0 & \frac{1}{c'} \\ 0 & \frac{1}{c'} & 0 & 0 \end{bmatrix}$$

W_c, is the upper triangular Toeplitz matrix, and it is formed by characteristic equation of matrix A : $|sI - A| = s^4 + a_3 s^3 + a_2 s^2 + a_1 s + a_0 = s^4$. So,

$$W_c = \begin{bmatrix} 1 & a_3 & a_2 & a_1 \\ 0 & 1 & a_3 & a_2 \\ 0 & 0 & 1 & a_3 \\ 0 & 0 & 0 & 1 \end{bmatrix} = \begin{bmatrix} 1 & 0 & 0 & 0 \\ 0 & 1 & 0 & 0 \\ 0 & 0 & 1 & 0 \\ 0 & 0 & 0 & 1 \end{bmatrix} = I$$

As known,

$$\dot{\xi} = A\xi + B\tau \tag{3.22}$$

Substituting the Eq. (3.39) into Eq. (3.40), we obtain

$$\dot{\zeta} = [T_{cc}]^{-1} A [T_{cc}]\zeta + [T_{cc}]^{-1} B\tau \tag{3.23}$$

Hence, the controllable canonical form is given by

$$\begin{bmatrix} \dot{\zeta}_1 \\ \dot{\zeta}_2 \\ \dot{\zeta}_3 \\ \dot{\zeta}_4 \end{bmatrix} = \begin{bmatrix} 0 & 0 & 0 & 0 \\ 0 & 0 & 0 & 0 \\ 1 & 0 & 0 & 0 \\ 0 & 1 & 0 & 0 \end{bmatrix} \begin{bmatrix} \zeta_1 \\ \zeta_2 \\ \zeta_3 \\ \zeta_4 \end{bmatrix} + \begin{bmatrix} 1 & 0 \\ 0 & 1 \\ 0 & 0 \\ 0 & 0 \end{bmatrix} \begin{bmatrix} \tau_{1without} \\ \tau_{2without} \end{bmatrix} \tag{3.24}$$

and the state vector

$$\begin{bmatrix} \zeta_1 \\ \zeta_2 \\ \zeta_3 \\ \zeta_4 \end{bmatrix} = [T_{cc}]^{-1}\xi(t) = \begin{bmatrix} d'\dot{\theta}_1 \\ c'\dot{\theta}_2 \\ d'\theta_1 \\ c'\theta_2 \end{bmatrix} \tag{3.25}$$

It is obviously that, the system that described by state Eq. (3.42) can be divided into two independent subsystems:

$$\begin{cases} \begin{bmatrix} \dot{\zeta}_1 \\ \dot{\zeta}_3 \end{bmatrix} = \begin{bmatrix} 0 & 0 \\ 1 & 0 \end{bmatrix} \begin{bmatrix} \zeta_1 \\ \zeta_3 \end{bmatrix} + \begin{bmatrix} 1 \\ 0 \end{bmatrix} \tau_{1without} \\ \begin{bmatrix} \dot{\zeta}_2 \\ \dot{\zeta}_4 \end{bmatrix} = \begin{bmatrix} 0 & 0 \\ 1 & 0 \end{bmatrix} \begin{bmatrix} \zeta_2 \\ \zeta_4 \end{bmatrix} + \begin{bmatrix} 1 \\ 0 \end{bmatrix} \tau_{2without} \end{cases} \tag{3.26}$$

Hence, when taking into account the feedback control, the closed-loop can be added firstly on each simple subsystem, then combine them together for the whole system. Obviously, both of these two subsystems are simple double integrator structures. Hence, the linear control method can be employed.

3.2.1 Command of the First Double Integrator

According to the first double integrator, we can deduce:

$$\begin{cases} \dot{\zeta}_1 = \tau_{1without} \\ \dot{\zeta}_3 = \zeta_1 \end{cases} \tag{3.27}$$

Assuming that $\begin{cases} \rho = \zeta_3 \\ \dot{\rho} = \zeta_1 \end{cases}$

Then the state equation of the double integrator is given by

$$\begin{bmatrix} \dot{\rho} \\ \ddot{\rho} \end{bmatrix} = \underbrace{\begin{bmatrix} 0 & 1 \\ 0 & 0 \end{bmatrix}}_{A_1} \begin{bmatrix} \rho \\ \dot{\rho} \end{bmatrix} + \underbrace{\begin{bmatrix} 0 \\ 1 \end{bmatrix}}_{B_1} \tau_{1without} \tag{3.28}$$

Obviously, this is a simple double integrator structure. For the closed-loop control, the control law can be written as

$$\tau_{1without} = \mu_1 - g_{11}\dot{\rho} - g_{12}\rho \tag{3.29}$$

Substituting this control law in the Eq. (3.28), then the state equation of the double integrator can be rewritten as

$$\begin{bmatrix} \dot{\rho} \\ \ddot{\rho} \end{bmatrix} = \begin{bmatrix} 0 & 1 \\ -g_{12} & -g_{11} \end{bmatrix} \begin{bmatrix} \rho \\ \dot{\rho} \end{bmatrix} + \begin{bmatrix} 0 \\ 1 \end{bmatrix} \mu_1 \tag{3.30}$$

Assuming that

$$\mu_1 = \omega_1 + g_{11}\dot{\rho} + g_{12}\rho \tag{3.31}$$

then

$$\begin{bmatrix} \dot{\rho} \\ \ddot{\rho} \end{bmatrix} = \begin{bmatrix} 0 & 1 \\ 0 & 0 \end{bmatrix} \begin{bmatrix} \rho \\ \dot{\rho} \end{bmatrix} + \begin{bmatrix} 0 \\ 1 \end{bmatrix} \omega_1 \tag{3.32}$$

hence, $\omega_1 = \ddot{\rho}$. It can be found that

$$\mu_1 = \ddot{\rho} + g_{11}\dot{\rho} + g_{12}\rho \tag{3.33}$$

It is concluded that a balance position of the double integrator is given by the conditions about ρ and its continual derivatives.

Hence, if a trajectory ρ_R which is secondary derivable over a range [0, T] is given, we can deduce:

$$\mu_{1R} = \ddot{\rho}_R + g_{11}\dot{\rho}_R + g_{12}\rho_R \tag{3.34}$$

Finally, the control law of the closed-loop double integrator is

$$\tau_{1without} = \ddot{\rho}_R - g_{11}[\dot{\rho} - \dot{\rho}_R] - g_{12}[\rho - \rho_R] \tag{3.35}$$

As known, $\rho = d'\theta_1$, the control law can be rewritten as

$$\tau_{1without} = d'\ddot{\theta}_{1R} - g_{11}d'[\dot{\theta}_1 - \dot{\theta}_{1R}] - g_{12}d'[\theta_1 - \theta_{1R}] \tag{3.36}$$

3.2.2 Determination of the Feedback Parameters for the First Double Integrator

In order to realize the control law: $\tau_1(t) = -G_1[\rho(t) \quad \dot{\rho}(t)]^{\mathrm{T}}$, the matrix of state feedback parameter $G_1 = [g_{12} \quad g_{11}]$ should be obtained.

It is calculated by minimizing the index of performance J below:

$$J = \int_0^\infty \left[[\rho(t) \quad \dot{\rho}(t)] Q_1 \begin{bmatrix} \rho(t) \\ \dot{\rho}(t) \end{bmatrix} + \tau_1^T(t) R_1 \tau_1(t) \right] dt \tag{3.37}$$

where

$$R_1 = 1 > 0$$

$$Q_1 = \left[T_{p1} \left(\int_0^{T_{p1}} e^{A_1 \tau} B_1 B_1^T e^{A_1^T \tau} d\tau \right) \right]^{-1} = \begin{bmatrix} \frac{12}{T_{p1}^4} & -\frac{6}{T_{p1}^3} \\ -\frac{6}{T_{p1}^3} & \frac{4}{T_{p1}^2} \end{bmatrix} > 0$$

Assuming that the symmetric matrix Σ_1 is the solution of Riccati equation

$$A_1^T \Sigma_1 + \Sigma_1 A_1 - \Sigma_1 B_1 R_1^{-1} B_1^T \Sigma_1 + Q_1 = 0 \tag{3.38}$$

where $\Sigma_1 = \begin{bmatrix} \sigma_1 & \sigma_2 \\ \sigma_2 & \sigma_3 \end{bmatrix}$ must be positive-definite, i.e. $\begin{cases} \sigma_1 > 0 \\ \sigma_1\sigma_3 - \sigma_2^2 > 0 \end{cases}$

thus $\begin{cases} \sigma_1 = \sigma_2\sigma_3 + \frac{6}{T_{p1}^3} > 0; \\ \sigma_2^2 = \frac{12}{T_{p1}^4} \Rightarrow \sigma_2 = \pm\frac{2\sqrt{3}}{T_{p1}^2}; \\ \sigma_3^2 = 2\sigma_2 + \frac{4}{T_{p1}^2} > 0. \end{cases}$

we obtain

$$\begin{cases} 2\sigma_2 + \frac{4}{T_{p1}^2} > 0 \Rightarrow \sigma_2 = \frac{2\sqrt{3}}{T_{p1}^2} \Rightarrow \sigma_3^2 = \frac{4(1+\sqrt{3})}{T_{p1}^2} \\ \sigma_2\sigma_3 + \frac{6}{T_{p1}^3} > 0 \Rightarrow \sigma_3 = \frac{2\sqrt{1+\sqrt{3}}}{T_{p1}} \\ \sigma_1 = \sigma_2\sigma_3 + \frac{6}{T_{p1}^3} = \frac{6+4\sqrt{3}\sqrt{1+\sqrt{3}}}{T_{p1}^3} \end{cases} \tag{3.39}$$

Hence, the expression of matrix Σ_1 is

$$\Sigma_1 = \begin{bmatrix} \frac{6+4\sqrt{3}\sqrt{1+\sqrt{3}}}{T_{p1}^3} & \frac{2\sqrt{3}}{T_{p1}^2} \\ \frac{2\sqrt{3}}{T_{p1}^2} & \frac{2\sqrt{1+\sqrt{3}}}{T_{p1}} \end{bmatrix} \tag{3.40}$$

As known, the matrix G_1 is noted as

$$G_1 = R_1^{-1} B_1^T \Sigma_1 \tag{3.41}$$

Then,

$$G_1 = \begin{bmatrix} \frac{2\sqrt{3}}{T_{p1}^2} & \frac{2\sqrt{1+\sqrt{3}}}{T_{p1}} \end{bmatrix} \tag{3.42}$$

As result, we obtain $g_{12} = \frac{2\sqrt{3}}{T_{p1}^2}$ and $g_{11} = \frac{2\sqrt{1+\sqrt{3}}}{T_{p1}}$.

The state model of the closed-loop system is written as follows

$$\begin{bmatrix} \dot{\rho} \\ \ddot{\rho} \end{bmatrix} = [A_1 - B_1 G_1] \begin{bmatrix} \rho \\ \dot{\rho} \end{bmatrix} = \begin{bmatrix} 0 & 1 \\ -\frac{2\sqrt{3}}{T_{p1}^2} & -\frac{2\sqrt{1+\sqrt{3}}}{T_{p1}} \end{bmatrix} \begin{bmatrix} \rho \\ \dot{\rho} \end{bmatrix} \tag{3.43}$$

The characteristic polynomial P(s) is given by

$$P(s) = |sI - A_1 + B_1 G_1| = s^2 + \frac{2\sqrt{1+\sqrt{3}}}{T_{p1}} s + \frac{2\sqrt{3}}{T_{p1}^2} \tag{3.44}$$

If $P(s) = s^2 + 2\zeta\omega_n s + \omega_n^2$, we obtain $\omega_n = \frac{\sqrt{2\sqrt{3}}}{T_{p1}}$ and $\zeta = \frac{\sqrt{1+\sqrt{3}}}{\sqrt{2\sqrt{3}}} \approx 0.89$.

According to these results, the control law and feedback parameters of the second double integrator can be written as

$$\tau_{2without} = c'\ddot{\theta}_{2R} - g_{21} c'[\dot{\theta}_2 - \dot{\theta}_{2R}] - g_{22} c'[\theta_2 - \theta_{2R}] \tag{3.45}$$

where, $g_{22} = \frac{2\sqrt{3}}{T_{p2}^2}$ and $g_{21} = \frac{2\sqrt{1+\sqrt{3}}}{T_{p2}}$.

3.3 Dynamic Analysis Taking into Account the Payload

Based on the assumption above, we consider the link 2 and the payload as one object. Then,

$$M_2 = m_2 + m_p \tag{3.46}$$

$$L_{BS2r} = \frac{m_2}{M_2} L_{BS2} + \frac{m_p}{M_2} L_{BP} \tag{3.47}$$

$$I_{2+p} = I_2 + m_2(L_{BS2} - L_{BS2r})^2 + m_p(L_2 - L_{BS2r})^2 \tag{3.48}$$

where, M_2 is the total mass of link 2 and the payload; L_{BS2r} is the distance between the mass center of M_2 and the joint center B; L_{BP} is the distance between the joint center B and the end-effector; I_{2+p} is the total moment of inertia of link 2 and the payload.

Thus, according to the Lagrangian dynamics, the dynamic equations with payload are:

$$\begin{aligned}
\tau_{1with} &= \frac{d}{dt}\left(\frac{\partial L}{\partial \dot{\theta}_1}\right) - \frac{\partial L}{\partial \theta_1} \\
&= \ddot{\theta}_1 [m_1 L_{AS1}^2 + I_{S1} + M_2 L_1^2 + M_2 L_{BS2r}^2 + 2M_2 L_1 L_{BS2r}\cos(\theta_2) + I_{2+P} \\
&\quad + m_3 L_1^2 + m_3 L_3^2 + m_3 L_{CS3}^2 - 2m_3 L_1 L_3 \cos(\theta_2) - 2m_3 L_1 L_{CS3}\cos(\theta_2) \\
&\quad + 2m_3 L_3 L_{CS3}\cos(2\theta_2) + I_{S3} + m_4(2L_3\cos(\theta_2) - L_1)^2 + I_{S4}] \\
&\quad + \ddot{\theta}_2 [M_2 L_{BS2r}^2 + M_2 L_1 L_{BS2r}\cos(\theta_2) + I_{2+P} + m_3 L_3^2 - m_3 L_{CS3}^2 \\
&\quad - m_3 L_1 L_3 \cos(\theta_2) + m_3 L_1 L_{CS3}\cos(\theta_2) - I_{S3}] \\
&\quad + \dot{\theta}_1\dot{\theta}_2[-2M_2 L_1 L_{BS2r}\sin(\theta_2) + 2m_3 L_1 L_3 \sin(\theta_2) + 2m_3 L_1 L_{CS3}\sin(\theta_2) \\
&\quad - 4m_3 L_3 L_{CS3}\sin(2\theta_2) - 4m_4 L_3(2L_3\cos(\theta_2) - L_1)\sin(\theta_2)] \\
&\quad + \dot{\theta}_2^2[-M_2 L_1 L_{BS2r}\sin(\theta_2) + m_3 L_1 L_3 \sin(\theta_2) - m_3 L_1 L_{CS3}\sin(\theta_2)]
\end{aligned} \tag{3.49}$$

$$\begin{aligned}
\tau_{2with} &= \frac{d}{dt}\left(\frac{\partial L}{\partial \dot{\theta}_2}\right) - \frac{\partial L}{\partial \theta_2} \\
&= \ddot{\theta}_1 [M_2 L_{BS2r}^2 + 2M_2 L_1 L_{BS2r}\cos(\theta_2) + I_{2+p} + m_3 L_3^2 - m_3 L_{CS3}^2 \\
&\quad - m_3 L_1 L_3 \cos(\theta_2) + m_3 L_1 L_{CS3}\cos(\theta_2) - I_{S3}] \\
&\quad + \ddot{\theta}_2 [M_2 L_{BS2r}^2 + I_{2+p} + m_3 L_3^2 + m_3 L_{CS3}^2 - 2m_3 L_3 L_{CS3}\cos(2\theta_2) + I_{S3} \\
&\quad + 4m_4 L_3^2 \sin^2(\theta_2)] \\
&\quad + \dot{\theta}_1^2[M_2 L_1 L_{BS2r}\sin(\theta_2) - m_3 L_1 L_3 \sin(\theta_2) - m_3 L_1 L_{CS3}\sin(\theta_2) \\
&\quad + 2m_3 L_3 L_{CS3}\sin(2\theta_2) + 2m_4 L_3(2L_3\cos(\theta_2) - L_1)\sin(\theta_2)] \\
&\quad + \dot{\theta}_2^2[2m_3 L_3 L_{CS3}\sin(2\theta_2) - 2m_4 L_3^2 \sin(2\theta_2)]
\end{aligned} \tag{3.50}$$

where, τ_{1with} and τ_{2with} are the output torque values of the first actuator and the second actuator respectively when taking into account the payload on the end-effector of the serial manipulator.

From Eqs. (3.15), (3.16) and (3.47), we can obtain that

$$L_{CS3} = -\frac{m_4}{m_3} L_3 \tag{3.51}$$

$$L_{BS2r} = \frac{(m_3+m_4)}{M_2}L_3 + \frac{m_p}{M_2}L_{BP} \tag{3.52}$$

Substituting these two equations into the dynamic Eqs. (3.49) and (3.50), then they can be simplified as

$$\begin{aligned}\tau_{1with} = {} & \ddot{\theta}_1[I_{S1} + I_{2+p} + I_{S3} + I_{S4} + m_1L_{AS1}^2 + M_2L_1^2 + m_3L_1^2 + m_4L_1^2 \\ & + \frac{[(m_3+m_4)L_3 + m_pL_{BP}]^2}{M_2} + \frac{(m_3+m_4)^2}{m_3}L_3^2 + 2m_pL_1L_{BP}\cos(\theta_2)] \\ & + \ddot{\theta}_2[I_{2+p} - I_{S3} + \frac{[(m_3+m_4)L_3 + m_pL_2]^2}{M_2} + \frac{m_3^2 - m_4^2}{m_3}L_3^2] \\ & + \ddot{\theta}_2[m_pL_1L_{BP}\cos(\theta_2)] + \dot{\theta}_1\dot{\theta}_2[-2m_pL_1L_{BP}\sin(\theta_2)] \\ & + \dot{\theta}_2^2[-m_pL_1L_{BP}\sin(\theta_2)]\end{aligned} \tag{3.53}$$

$$\begin{aligned}\tau_{2with} = {} & \ddot{\theta}_1[I_{2+p} - I_{S3} + \frac{[(m_3+m_4)L_3 + m_pL_{BP}]^2}{M_2} + \frac{m_3^2 - m_4^2}{m_3}L_3^2] \\ & + \ddot{\theta}_2[I_{2+p} + I_{S3} + \frac{[(m_3+m_4)L_3 + m_pL_{BP}]^2}{M_2} + \frac{(m_3+m_4)^2}{m_3}L_3^2] \\ & + \ddot{\theta}_1[m_pL_1L_{BP}\cos(\theta_2)] + \dot{\theta}_1^2[m_pL_{BP}]L_1\sin(\theta_2)\end{aligned} \tag{3.54}$$

Finally, replace the parameter I_{S3} and I_{2+p} by the Eqs. (3.18) and (3.48), the dynamic equation with payload can be rewritten as

$$\begin{cases}\tau_{1with} = \ddot{\theta}_1[I_{S1} + 2I_{S2} + I_{S4} + m_1L_{AS1}^2 + (m_2+m_3+m_4)L_1^2 + 2\frac{(m_3+m_4)(m_2+m_3+m_4)}{m_2} \\ \quad L_3^2] + \ddot{\theta}_1[m_pL_1^2] + (\ddot{\theta}_1 + \ddot{\theta}_2)[\frac{[m_pL_{BP} + (m_3+m_4)L_3]^2}{M_2} - \frac{(m_3+m_4)^2}{m_2}L_3^2] \\ \quad + \ddot{\theta}_1[2m_pL_1L_{BP}]\cos\theta_2 + \ddot{\theta}_2[m_pL_1L_{BP}]\cos\theta_2 + \dot{\theta}_1\dot{\theta}_2[-2m_pL_1L_{BP}]\sin\theta_2 \\ \quad + \dot{\theta}_2^2[-m_pL_1L_{BP}]\sin\theta_2 \\ \tau_{2with} = 2\ddot{\theta}_2[I_{S2} + \frac{(m_3+m_4)(m_2+m_3+m_4)}{m_2}L_3^2] \\ \quad + (\ddot{\theta}_1 + \ddot{\theta}_2)[\frac{[m_pL_{BP} + (m_3+m_4)L_3]^2}{M_2} - \frac{(m_3+m_4)^2}{m_2}L_3^2] \\ \quad + \ddot{\theta}_1[m_pL_1L_{BP}]\cos\theta_2 + \dot{\theta}_1^2[m_pL_1L_{BP}]\sin\theta_2\end{cases} \tag{3.55}$$

Compare Eq. (3.55) with Eq. (3.19), the torques related to the payload can be found.

$$\begin{cases} \Delta\tau_1 = \ddot{\theta}_1[m_pL_1^2] + (\ddot{\theta}_1 + \ddot{\theta}_2)[\frac{[m_pL_{BP} + (m_3 + m_4)L_3]^2}{M_2} - \frac{(m_3 + m_4)^2}{m_2}L_3^2] \\ \quad + \ddot{\theta}_1[2m_pL_1L_{BP}]\cos\theta_2 + \ddot{\theta}_2[m_pL_1L_{BP}]\cos\theta_2 + \dot{\theta}_1\dot{\theta}_2[-2m_pL_1L_{BP}]\sin\theta_2 \\ \quad + \dot{\theta}_2^2[-m_pL_1L_{BP}]\sin\theta_2 \\ \Delta\tau_2 = (\ddot{\theta}_1 + \ddot{\theta}_2)[\frac{[m_pL_{BP} + (m_3 + m_4)L_3]^2}{M_2} - \frac{(m_3 + m_4)^2}{m_2}L_3^2] \\ \quad + \ddot{\theta}_1[m_pL_1L_{BP}]\cos\theta_2 + \dot{\theta}_1^2[m_pL_1L_{BP}]\sin\theta_2 \end{cases} \tag{3.56}$$

Hence, the dynamic equation with payload can be rewritten as

$$\begin{cases} \tau_{1with} = \tau_{1without} + \Delta\tau_1 \\ \tau_{2with} = \tau_{2without} + \Delta\tau_2 \end{cases} \tag{3.57}$$

As noted in the previous chapter, this part is referred as payload compensation. Extracting the payload compensation from the dynamic equation of the model taking into account the payload, then the model can be treated still as a dynamic decoupling model. Also, no matter there is payload or not, the simple linear control law can always be used.

3.4 Illustrative Example with the SIMULINK Block of MATLAB

For verifying the performance of this proposed dynamic decoupling method, a simulation model will be built in the SIMULINK block of MATLAB. We note that

$$\begin{aligned} a_0 &= [m_1L_{AS1}^2 + I_{S1} + M_2L_1^2 + M_2L_{BS2r}^2 + I_{2+p} + m_3L_1^2 + m_3L_3^2 \\ &\quad + m_3L_{CS3}^2 + I_{S3} + I_{S4} + 4m_4L_3^2 + m_4L_1^2 + 2m_3L_3L_{CS3}] \\ a_1 &= [m_3L_3L_{CS3} + m_4L_3^2] \\ a_2 &= [M_2L_1L_{BS2r} - m_3L_1L_3] \\ a_3 &= [M_2L_{BS2r}^2 + I_{2+p} + m_3L_3^2] \\ a_4 &= [m_3L_{CS3}^2 + I_3] \\ a_5 &= [m_3L_1L_3] \\ a_6 &= [m_4L_1L_3] \\ a_7 &= [m_3L_3L_{CS3}] \end{aligned} \tag{3.58}$$

then, the dynamic Eqs. (3.49) and (3.50) can be rewritten as

$$\begin{aligned}\tau_1 &= a_0\ddot{\theta}_1 + \ddot{\theta}_1[2a_2 - 2a_5 - 4a_6]\cos(\theta_2) + \ddot{\theta}_1[-4a_1]\sin^2(\theta_2) \\ &+ \ddot{\theta}_2[a_3 - a_4] + \ddot{\theta}_2[a_2 + a_5]\cos(\theta_2) + \dot{\theta}_1\dot{\theta}_2[-2a_2 + 2a_5 + 4a_6]\sin(\theta_2) \\ &+ \dot{\theta}_1\dot{\theta}_2[-8a_1]\sin(\theta_2)\cos(\theta_2) + \dot{\theta}_2^2[-a_2 - a_5]\sin(\theta_2)\end{aligned} \tag{3.59}$$

$$\begin{aligned}\tau_2 &= \ddot{\theta}_1[a_3 - a_4] + \ddot{\theta}_1[a_2 + a_5]\cos(\theta_2) \\ &+ \ddot{\theta}_2[a_3 + a_4 - 2a_7] + \ddot{\theta}_2[4a_1]\sin^2(\theta_2) \\ &+ \dot{\theta}_1^2[a_2 - a_5 - 2a_6]\sin(\theta_2) + \dot{\theta}_1^2[4a_1]\sin(\theta_2)\cos(\theta_2) \\ &+ \dot{\theta}_2^2[4a_1]\sin(\theta_2)\cos(\theta_2)\end{aligned} \tag{3.60}$$

According to the form $A\underline{\ddot{q}} + C\underline{\dot{q}} = \underline{\tau}$, we can obtain the dynamic model equation as

$$\begin{aligned}\begin{bmatrix}\tau_1 \\ \tau_2\end{bmatrix} &= \underbrace{\begin{bmatrix} a_0 + 2\alpha_1(\theta_2) - 4\beta_1(\theta_2) & a_3 - a_4 + \gamma_1(\theta_2) \\ a_3 - a_4 + \gamma_1(\theta_2) & a_3 + a_4 - 2\phi_7 + 4\beta_1(\theta_2)\end{bmatrix}}_{A}\underbrace{\begin{bmatrix}\ddot{\theta}_1 \\ \ddot{\theta}_2\end{bmatrix}}_{\ddot{q}} \\ &+ \underbrace{\begin{bmatrix}\dot{\theta}_2[-2\alpha_2(\theta_2) - 8\beta_2(\theta_2)] & \dot{\theta}_2[-\gamma_2(\theta_2)] \\ \dot{\theta}_1[\alpha_2(\theta_2) + 4\beta_2(\theta_2)] & \dot{\theta}_2[4\beta_2(\theta_2)]\end{bmatrix}}_{C}\underbrace{\begin{bmatrix}\dot{\theta}_1 \\ \dot{\theta}_2\end{bmatrix}}_{\dot{q}}\end{aligned} \tag{3.61}$$

where

$$\begin{aligned}\alpha_1(\theta_2) &= [a_2 - a_5 - a_6]\cos(\theta_2) \\ \alpha_2(\theta_2) &= [a_2 - a_5 - a_6]\sin(\theta_2) \\ \beta_1(\theta_2) &= a_1\sin^2(\theta_2) \\ \beta_2(\theta_2) &= a_1\sin(\theta_2)\cos(\theta_2) \\ \gamma_1(\theta_2) &= [a_2 + a_5]\cos(\theta_2) \\ \gamma_2(\theta_2) &= [a_2 + a_5]\sin(\theta_2)\end{aligned}$$

As the matrix A is positive and invertible, we deduce

$$A^{-1} = \frac{1}{\Delta(\theta_2)}\begin{bmatrix} a_3 + a_4 - 2a_7 + 4\beta_1(\theta_2) & -a_3 + a_4 - \gamma_1(\theta_2) \\ -a_3 + a_4 - \gamma_1(\theta_2) & a_0 + 2\alpha_1(\theta_2) - 4\beta_1(\theta_2)\end{bmatrix} \tag{3.62}$$

where

$$\Delta(\theta_2) = [a_0 + 2\alpha_1(\theta_2) - 4\beta_1(\theta_2)][a_3 + a_4 - 2a_7 + 4\beta_1(\theta_2)] - [a_3 - a_4 + \gamma_1(\theta_2)]^2$$

Finally, the serial planar manipulator is modeled in MATLAB by using the following equations:

$$\begin{aligned}
\ddot{\theta}_1 = & \frac{a_3 + a_4 - 2a_7 + 4\beta_1(\theta_2)}{\Delta(\theta_2)}\tau_1 + \frac{-a_3 + a_4 - \gamma_1(\theta_2)}{\Delta(\theta_2)}\tau_2 \\
& + \frac{2[a_3 + a_4 - 2a_7 + 4\beta_1(\theta_2)][\alpha_2(\theta_2) + 4\beta_2(\theta_2)]}{\Delta(\theta_2)}\dot{\theta}_1\dot{\theta}_2 \\
& - \frac{[-a_3 + a_4 - \gamma_1(\theta_2)][\alpha_2(\theta_2) + 4\beta_2(\theta_2)]}{\Delta(\theta_2)}\dot{\theta}_1^2 \\
& + \frac{[a_3 + a_4 - 2a_7 + 4\beta_1(\theta_2)][\gamma_2(\theta_2)]}{\Delta(\theta_2)}\dot{\theta}_2^2 \\
& - \frac{4[-a_3 + a_4 - \gamma_1(\theta_2)][\beta_2(\theta_2)]}{\Delta(\theta_2)}\dot{\theta}_2^2
\end{aligned} \tag{3.63}$$

$$\begin{aligned}
\ddot{\theta}_2 = & \frac{-a_3 + a_4 - \gamma_1(\theta_2)}{\Delta(\theta_2)}\tau_1 + \frac{a_0 + 2\alpha_1(\theta_2) - 4\beta_1(\theta_2)}{\Delta(\theta_2)}\tau_2 \\
& + \frac{2[-a_3 + a_4 - \gamma_1(\theta_2)][\alpha_2(\theta_2) + 4\beta_2(\theta_2)]}{\Delta(\theta_2)}\dot{\theta}_1\dot{\theta}_2 \\
& - \frac{[a_0 + 2\alpha_1(\theta_2) - 4\beta_1(\theta_2)][\alpha_2(\theta_2) + 4\beta_2(\theta_2)]}{\Delta(\theta_2)}\dot{\theta}_1^2 \\
& + \frac{[-a_3 + a_4 - \gamma_1(\theta_2)][\gamma_2(\theta_2)]}{\Delta(\theta_2)}\dot{\theta}_2^2 \\
& - \frac{4[a_0 + 2\alpha_1(\theta_2) - 4\beta_1(\theta_2)][\beta_2(\theta_2)]}{\Delta(\theta_2)}\dot{\theta}_2^2
\end{aligned} \tag{3.64}$$

3.4.1 Simulation Model of Open-Loop Control System

First, the open-loop control system of the serial planar manipulator is built. It is constituted mainly by two parts, the controller part and the manipulator part. The controller part is described by the Eqs. (3.19) and (3.56). And the manipulator part is described by Eqs. (3.63) and (3.64), as shown in Fig. 3.2.

The parametric values of all the parts of the system are given in Table 3.1. It should be noted that the value of the length parameter L_{CS3} is negative. That is because a counterweight is added on the inverse extension line of link 3, when consider link 3 and the payload as one object, the total mass center moves along the inverse extension line of link 3 and finally located on the side of the joint C.

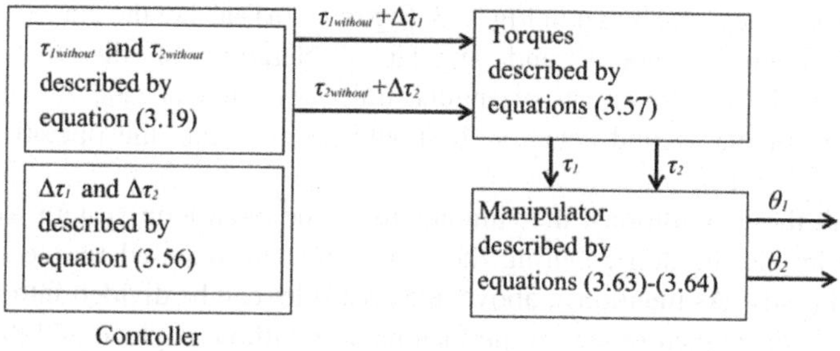

Fig. 3.2 The schema of the open-loop control system

Table 3.1 The parametric values of all the parts of the system

Mass (kg)		Length (m)		Moment of inertia ($kg \cdot m^2$)	
m_1	13.193	L_1/L_{AS1}	0.8/0.4	I_{S1}	0.77754
m_2	10.472	L_2/L_{BS2}	0.8/0.1738	I_{S2}	1.415
m_3	5.374	L_3/L_{CS3}	0.3/−0.0386	I_{S3}	2.2069
m_4	0.692	L_{BP}	0.5	I_{S4}	7.2×10^{-4}
m_p	5				

Table 3.2 The composite parametric values of the system

Mass (kg)		Length (m)		Moment of inertia ($kg \cdot m^2$)	
M_2	15.472	L_{BS2r}	0.2792	I_{2+p}	1.7751

As discussed above, in order to simplify the dynamic model, link 2 and the payload are treated as one object. The parametric values of the composite object are shown in Table 3.2.

The initial and final simulation angles are given in Table 3.3.

Different from the decoupled model that proposed in last chapter, there is no special requirement for choosing these angles, i.e. they can be arbitrary values. Here, for the desired trajectories, velocities and accelerations, the same expressions that described by Eqs. (2.29) and (2.30) are used.

Table 3.3 The initial and final values of the desired trajectories

Angle (°)		Velocity (m/s)		Acceleration (m/s^2)	
θ_{1I}	0	$\dot{\theta}_{1I}$	0	$\ddot{\theta}_{1I}$	0
θ_{1F}	40	$\dot{\theta}_{1F}$	0	$\ddot{\theta}_{1F}$	0
θ_{2I}	30	$\dot{\theta}_{2I}$	0	$\ddot{\theta}_{2I}$	0
θ_{2F}	80	$\dot{\theta}_{2F}$	0	$\ddot{\theta}_{2F}$	0

As a result, they are shown in Fig. 3.3. Figure 3.3a1–a3 are the curves of angular trajectory, angular velocity and angular acceleration of the first actuator. Figure 3.3b1–b3 are the curves of angular trajectory, angular velocity and angular acceleration of the second actuator. It should be note that, the operation time is assumed as T = 1 s.

With all these conditions, the simulations of the open-loop system with a payload of 5 kg can be done. During the simulation, the required torque values are calculated firstly. As mentioned above, the controller can be divided into two parts: the dynamic decoupled model of the manipulator without payload and the payload compensation shown in Fig. 3.2. With the payload compensation, the expressions of the torques are derived exactly by the model of the manipulator. Thus, the desired torque values can be obtained, shown as the solid lines in Fig. 3.4. Meanwhile, the desired trajectories are achieved, shown as the solid lines in Fig. 3.5.

When without the payload compensation, the torque expressions in the controller won't precisely reflect the model of the manipulator. Thus, under this situation, the torque values are different with the desired ones, shown as the dashed lines in Fig. 3.4. In other aspect, the payload compensation can be treated as the compensation of the inaccurate modeling of the manipulator. As known, the inaccurate modeling part is always exists more or less. Hence, without the payload compensation, a great difference will appear during the desired and the real trajectories (Fig. 3.5). The errors of angular displacements of link AB and BP are, respectively, 35.3 and 52.92%.

3.4.2 Simulation Model of Closed-Loop Control System

To deal with the unexpected disturbances or situations, such as the lack of the payload compensation, the closed-loop is added, shown in Fig. 3.6.

The feedback factors are already obtained above, according to the schema of the closed-loop control system, the closed-loop control law can be written as

$$\begin{aligned} \tau_1(t) &= \tau_{1without} + \Delta\tau_1 \\ &\quad - g_{11}d'[\dot{\theta}_1 - \dot{\theta}_{1R}] - g_{12}d'[\theta_1 - \theta_{1R}] \\ \tau_2(t) &= \tau_{2without} + \Delta\tau_2 \\ &\quad - g_{21}c'[\dot{\theta}_2 - \dot{\theta}_{2R}] - g_{22}c'[\theta_2 - \theta_{2R}] \end{aligned} \tag{3.65}$$

where, $\tau_{1without}$ and $\tau_{2without}$ are given in Eq. (3.19); $\Delta\tau_1$ and $\Delta\tau_2$ are given in Eq. (3.56).

As known, the feedback loop works only when there is difference between the desired and real trajectories. With the payload compensation, the desired torques

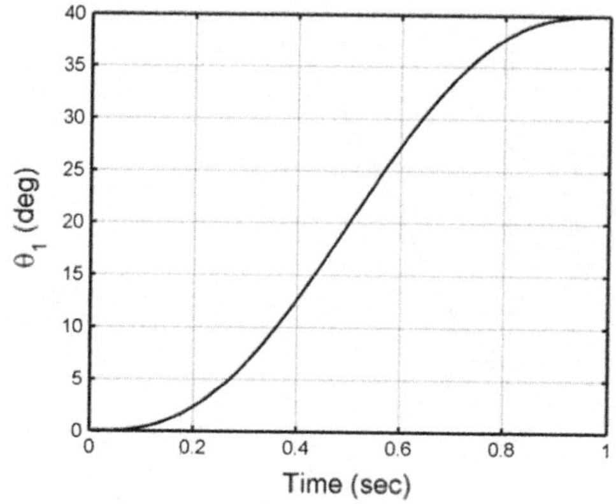

(a1)

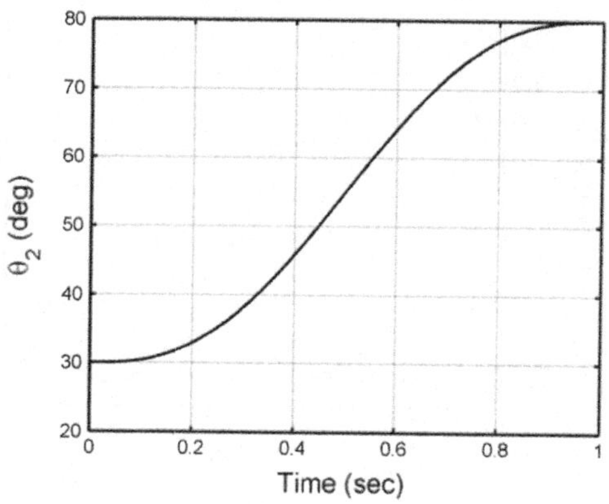

(b1)

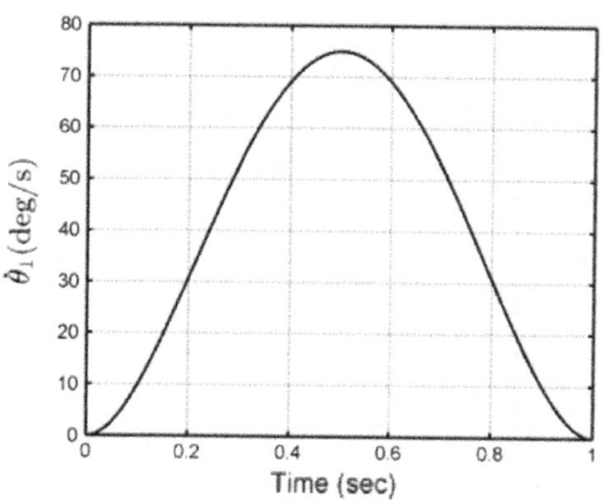

(a2)

Fig. 3.3 Desired trajectories, velocities and accelerations of the two actuators

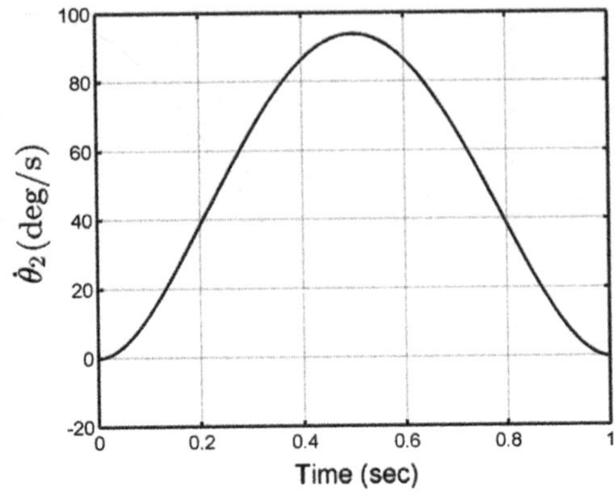

(b2)

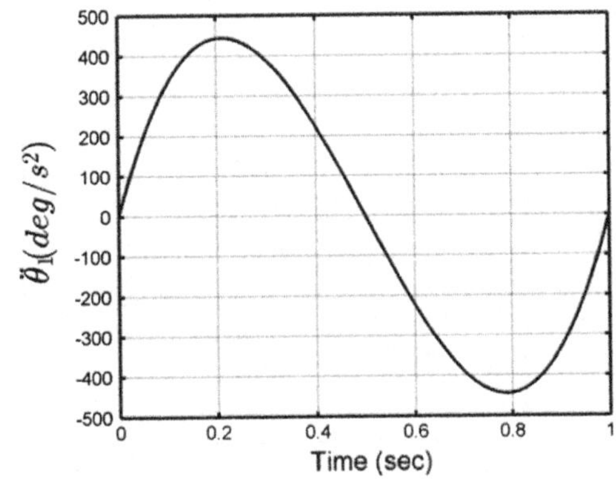

(a3)

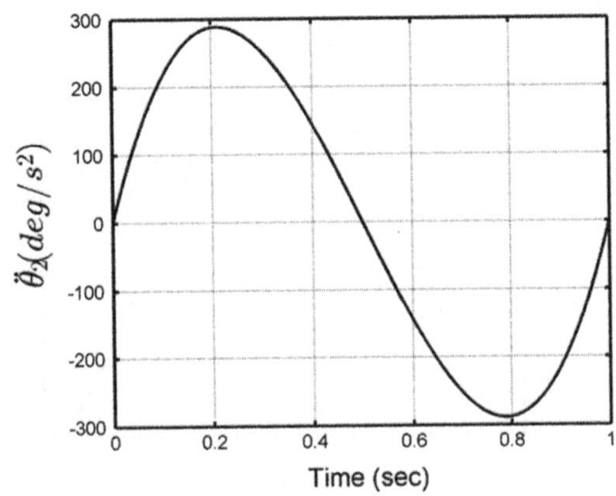

(b3)

Fig. 3.3 (continued)

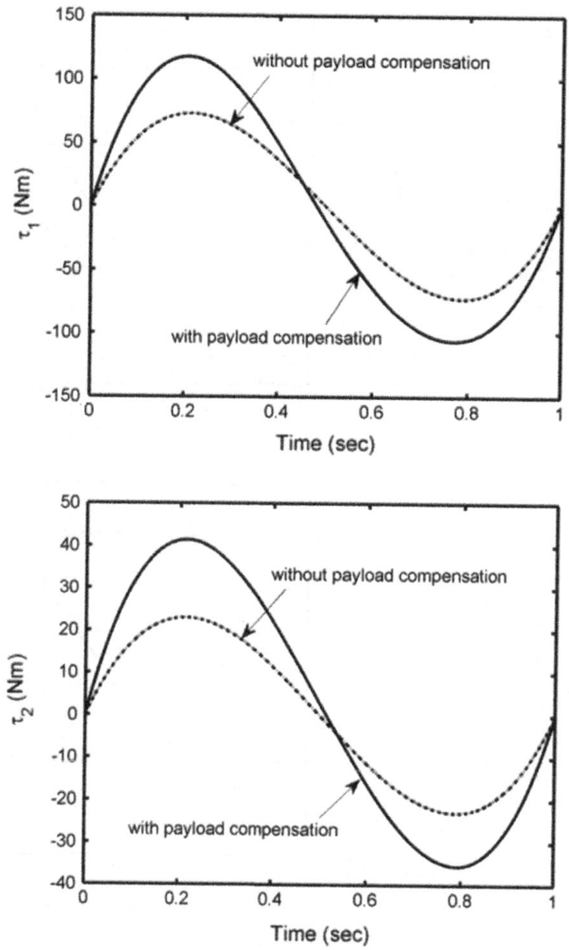

Fig. 3.4 Torques with payload compensation (solid line) and without it (dashed line) for the open-loop system

allows an exact reproduction of the desired manipulator motions, i.e. the tracking curves overlaps completely with the desired trajectories, shown as the solid lines in Figs. 3.7 and 3.8. If the payload compensation is absent in the controller, it will lead to inaccurate torques, hence the incorrect tracking trajectories will be obtained. However, such a tracking error can be reduced because of the self-correcting performance of the feedback loop. It is obvious that, the tracking curves without payload compensation (the dashed lines in Fig. 3.8) are closer with the desired trajectories than the ones in open-loop system (the dashed lines in Fig. 3.5). As a result, the errors of angular displacements of link AB and BP are, respectively, 2.68 and 2.96%. Obviously, they are greatly diminished.

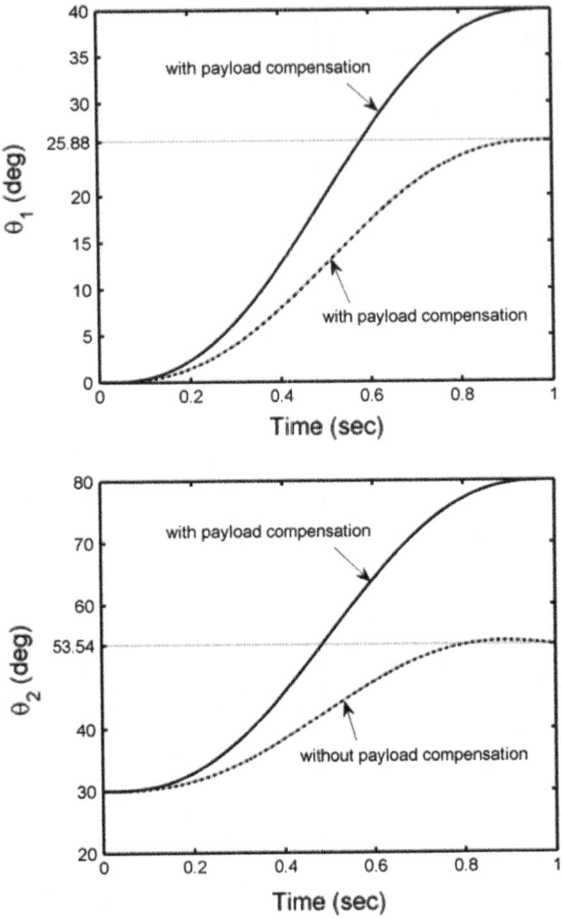

Fig. 3.5 Angular displacements of links with payload compensation (solid line) and without it (dashed line) for the open-loop system

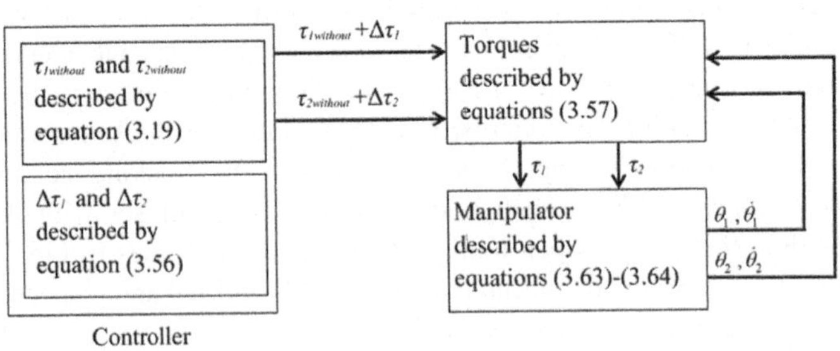

Fig. 3.6 The schema of the closed-loop control system

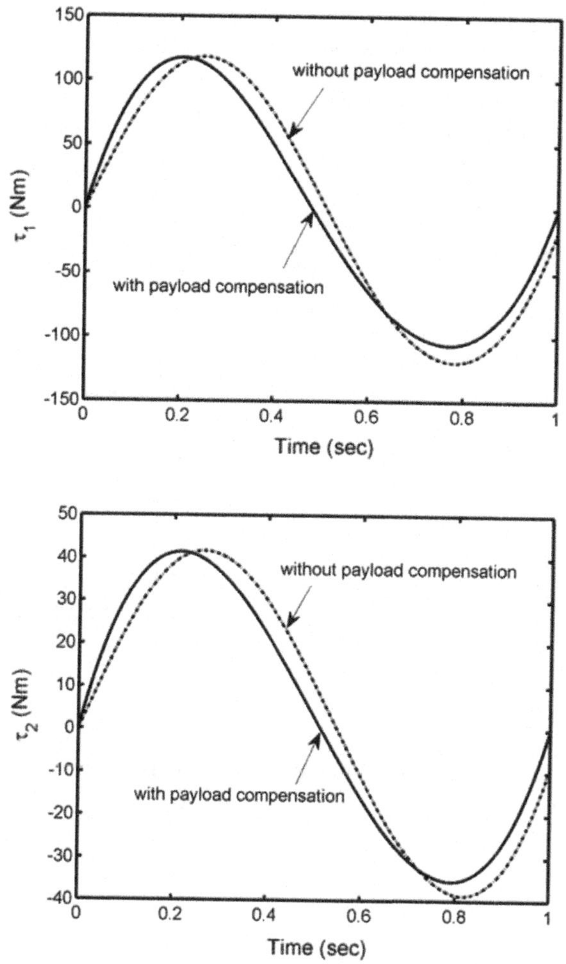

Fig. 3.7 Torques with payload compensation (solid line) and without it (dashed line) for the closed-loop system

3.5 Summary

This chapter proposed a new dynamic decoupling principle, which involves connecting to a serial manipulator a two-link group forming a Scott-Russell mechanism with the initial links of the manipulator.

It has been carried out in two steps. At first, the dynamic decoupling of the serial manipulator is accomplished via the Scott-Russell mechanism properties and optimal redistribution of masses. Thus, the modification of the mass redistribution allows one to transform the original nonlinear system model into a fully linear system without using the feedback linearization technique. However, as it mentioned above, the changing payload leads to the perturbation of the dynamic

Fig. 3.8 Angular displacements of links with load compensation (solid line) and without it (dashed line) for the closed-loop system

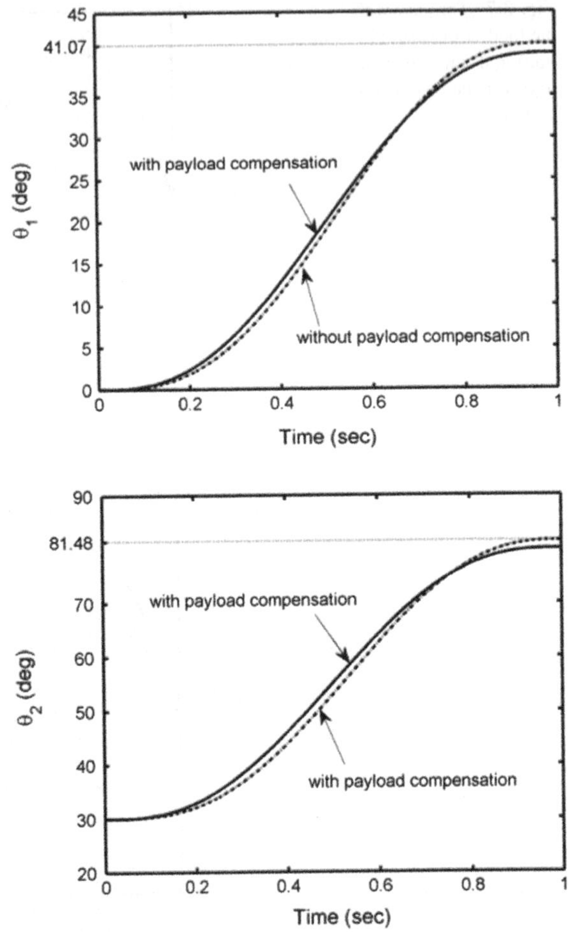

decoupling of the manipulator. To ensure decoupled dynamics of the manipulator for any payload, an optimal control technique has been applied.

The perturbation of required motions of the manipulator with payload compensation and without has been illustrated via simulations. Two kinds of simulations are carried out with open-loop control system which is a non-feedback system and closed-loop control system. The obtained results showed the efficiency of the proposed solution.

Finally, it should be noted that the developed approach of dynamic decoupling can also be applied to the design of planar serial manipulators with three degrees of freedom.

References

1. Freemantle W (1803) Straight-line linkage. British Patent 2741, 17 Nov 17 1803
2. Briot S, Bonev I, Gosselin C, Arakelian V (2009) Complete shaking force and shaking moment balancing of planar parallel manipulators with prismatic pairs. J Multi-body Dyn 223:43–52

Chapter 4
Tolerance Analysis of Serial Manipulators with Decoupled and Coupled Dynamics

J. Xu, V. Arakelian and J. P. Le Baron

Abstract This chapter deals with the robustness properties of serial manipulators with decoupled and coupled dynamics derived by tolerance analysis. After having introduced some performance indices of the manipulators, the tolerance capabilities of four manipulators are analyzed. In order to quantify the influencing degree, two kinds of the indices are defined. They are angular error and position error. Two kinds of simulation are designed here. The first kind of simulation is implemented by fixed parametric error. Then, the influencing degrees of all variables on the positioning accuracy of the manipulators are analyzed respectively. In order to obtain the models closer to the practical situation, the random parametric errors are introduced in the second kind of simulation. Furthermore, the parametric errors of all the variables are added at the same time during one simulation. The simulation results prove that the manipulators that decoupled by the mechatronic methods are more robust.

4.1 Performance Indices of the Manipulators

The performances of the manipulators such as the dexterity, load capacity, force transmission from the joint to the end-effector, and dynamic responsiveness etc. are investigated for optimize the design of the manipulators. Usually, in these researches, the performances are quantified by using some indices.

The dexterity indices for planar and spatial manipulators are presented in [1, 2] proposed the generalized inertia ellipsoid (GIE) as a quantitative method to measure the capability of changing end-effector's velocity in different directions for fixed kinetic energy. In the research of [3], the dynamic manipulability ellipsoid

J. Xu · V. Arakelian (✉) · J. P. Le Baron
Department of Mechanical and Control Systems Engineering, INSA,
20 avenue des Buttes de Coësmes, CS 70839, 35708 Rennes Cedex, France
e-mail: vigen.arakelyan@insa-rennes.fr

V. Arakelian · J. P. Le Baron
LS2 N/ECN, 1 rue de la Noë, BP 92101, 44321 Nantes Cedex 03, France

V. Arakelian (ed.), *Dynamic Decoupling of Robot Manipulators*, Mechanisms and Machine Science 56, https://doi.org/10.1007/978-3-319-74363-9_4

(DME) is introduced for measuring the ease of changing the end-effector's configuration by a set of joint torques with fixed magnitude. Both of the two indices (GIE and DME) are based on the relationship between the generalized inertia force of the end-effector and the generalized inertia torques of joints. The dynamic conditioning index (DCI) which is defined as the least-square difference between the generalized inertia matrix and an isotropic matrix is used to measure the dynamic performance of a manipulator [4].

Besides the analysis of the performances above, the error tolerance analysis is also important for a manipulator. For applications in remote and/or hazardous environments where repair is not possible, the fault tolerance of the manipulators is necessary. Cause once the components failure, it will result in a robot's joint becoming immobilized, i.e., a locked joint failure mode [5]. Hence, lots of researches such as to increase manipulator reliability [6] and to improve failure detection [7, 8] are proposed. However, these applications are extreme ones. In normal application of the manipulators where repair is possible, there is another performance referred as error tolerance corresponds to fault tolerance needs to be considered.

Positioning inaccuracy can stem from a number of sources such as the dimensional error of the components, the assembly error, the deflection error, the clearance in the kinematic pair, the elastic deformation error, the friction and wear error and the measurement and control error etc. In addition, during the actual operation, the variation of the payload, acceleration and deceleration of the manipulator may cause the geometric deviation and the movement deviation of the manipulator. All of these can affect the positioning accuracy of the serial manipulators.

The error of serial manipulator can be divided into systematic error and random error. System error is the error in the manufacturing and assembly process of the components. Usually, this kind of error is in the form of cumulative error, reverse error or periodic error. The error follows a certain mathematical model and can be compensated by control algorithm. Random error is the error that caused by uncertainty of the unpredictable disturbances, and therefore it is impossible to build a precise mathematical model for this error. The current effective way is to estimate the statistical processing of multiple measurements.

Considerable researches have been proposed for error analysis, error model derivation and calibration [9, 10, 11]. For the error model, some researches focus on the effects of manipulator joint errors [12, 13], as well as the effects of link dimensional errors [14, 15]. For instance, in the research of [16], two dimensional variations are discussed for a 2-DOF serial manipulator model.

4.2 The Dynamic Models of Manipulators for Tolerance Capability Comparison

The method mentioned in [16] is an efficient tolerance synthesis method. However, not just the length parameters, but also the parameters of mass and inertia are needed to be considered for tolerance analysis of the manipulators.

Hence, six main parameters of the serial manipulators are used here, they are:

- m_1, m_2: the mass parameters of the two main links 1 and 2 respectively;
- I_{S1}, I_{S2}: the inertia parameters of the two main links 1 and 2 respectively;
- L_{AS1}: the distance between the center of mass of link 1 and joint A;
- L_{BS2}: the distance between the center of mass of link 2 and joint B.

In addition, four serial planar manipulators with two degrees of freedom are considered. They are:

- manipulator_0: a serial planar manipulator with non-decoupled dynamics, shown in Fig. 4.1.
- manipulator_1: a serial planar manipulator with decoupled dynamics (Chap. 2), shown in Fig. 4.2.
- manipulator_2: the decoupled 2-DOF serial planar manipulator that proposed in Chap. 3, shown in Fig. 4.3.

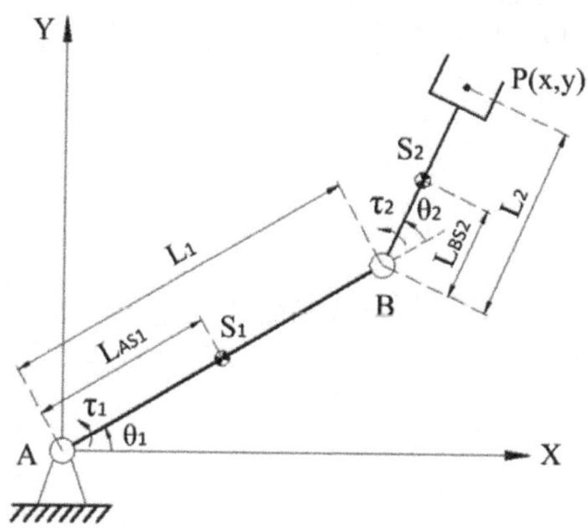

Fig. 4.1 The structure model of manipulator_0

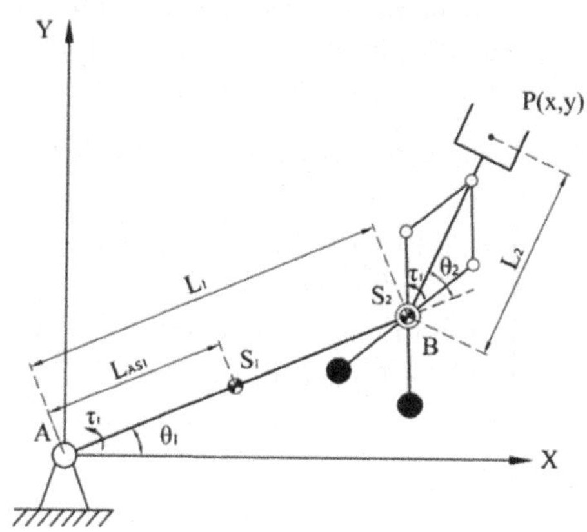

Fig. 4.2 The structure model of manipulator_1

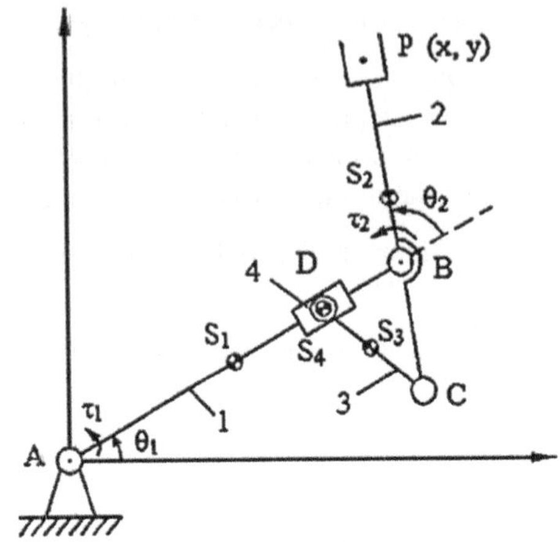

Fig. 4.3 The structure model of manipulator_2

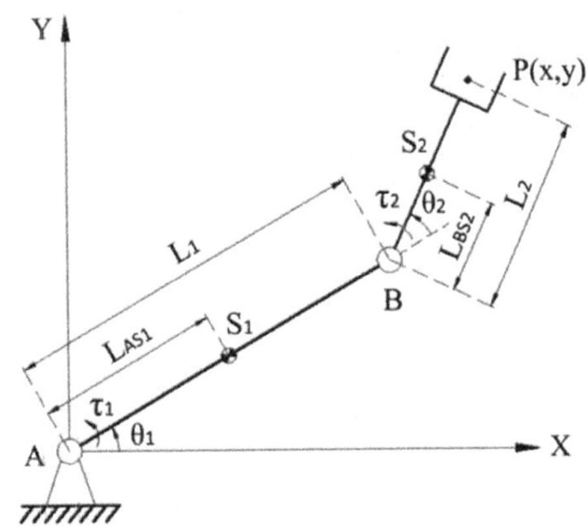

Fig. 4.4 The structure model of manipulator_3

- manipulator_3: a decoupled 2-DOF serial planar manipulator that decoupled by feedback linearization (Fig. 4.4). The control schema of inverse dynamics control is shown in Fig. 4.5.

The dynamic equation of manipulator_0 without payload can be expressed as

$$\begin{cases} \tau_1 = \ddot{\theta}_1 [a_{(0)} + b_{(0)} + 2m_2 L_1 L_{BS2} \cos\theta_2] + \ddot{\theta}_2 [a_{(0)} + m_2 L_1 L_{BS2} \cos\theta_2] \\ \qquad + \dot{\theta}_2^2 [-m_2 L_1 L_{BS2} \sin\theta_2] + \dot{\theta}_1 \dot{\theta}_2 [-2m_2 L_1 L_{BS2} \sin\theta_2] \\ \tau_2 = \ddot{\theta}_1 [a_{(0)} + m_2 L_1 L_{BS2} \cos\theta_2] + a_{(0)} \ddot{\theta}_2 \\ \qquad + \dot{\theta}_1^2 [m_2 L_1 L_{BS2} \sin\theta_2] \end{cases} \tag{4.1}$$

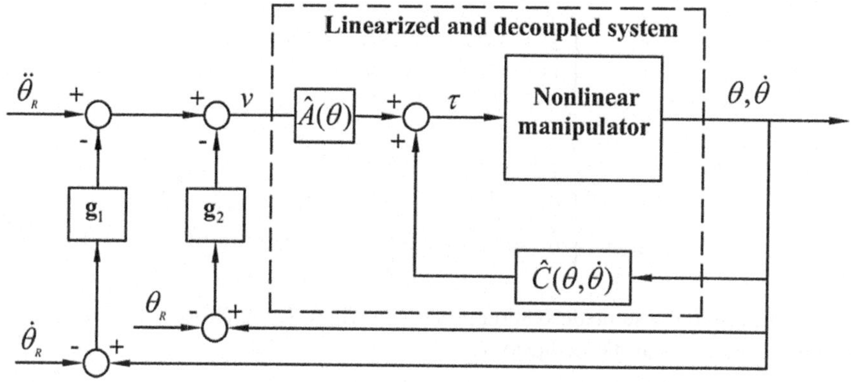

Fig. 4.5 The control schema of inverse dynamics control for manipulator_3

where,

$$a_{(0)} = m_2 L_{BS2}^2 + I_{S2}$$
$$b_{(0)} = m_1 L_{AS1}^2 + m_2 L_1^2 + I_{S1}$$

Thus the structure model of manipulator_0 for the simulation is given as

$$\begin{cases} \ddot{\theta}_1 = \frac{\gamma_2}{\Delta(\theta_2)}\tau_1 - \frac{[\gamma_2+\beta(\theta_2)]}{\Delta(\theta_2)}\tau_2 + \frac{\gamma_2\alpha(\theta_2)}{\Delta(\theta_2)}\dot{\theta}_1\dot{\theta}_2 \\ \quad + \frac{\alpha(\theta_2)[\gamma_2+\beta(\theta_2)]}{\Delta(\theta_2)}\dot{\theta}_1^2 + \frac{\gamma_2\alpha(\theta_2)}{\Delta(\theta_2)}(\dot{\theta}_1+\dot{\theta}_2)\dot{\theta}_2 \\ \ddot{\theta}_2 = -\frac{[\gamma_2+\beta(\theta_2)]}{\Delta(\theta_2)}\tau_1 + \frac{\psi(\theta_2)}{\Delta(\theta_2)}\tau_2 - \frac{\alpha(\theta_2)[\gamma_2+\beta(\theta_2)]}{\Delta(\theta_2)}\dot{\theta}_1\dot{\theta}_2 \\ \quad - \frac{\alpha(\theta_2)\psi(\theta_2)}{\Delta(\theta_2)}\dot{\theta}_1^2 - \frac{\alpha(\theta_2)[\gamma_2+\beta(\theta_2)]}{\Delta(\theta_2)}(\dot{\theta}_1+\dot{\theta}_2)\dot{\theta}_2 \end{cases} \tag{4.2}$$

where,

$$\alpha(\theta_2) = m_2 L_1 L_{BS2}\sin(\theta_2);$$
$$\beta(\theta_2) = m_2 L_1 L_{BS2}\cos(\theta_2);$$
$$\gamma_1 = I_{S1} + m_1 L_{AS1}^2;$$
$$\gamma_2 = I_{S2} + m_2 L_{BS2}^2;$$
$$\psi(\theta_2) = \gamma_1 + \gamma_2 + m_2 L_1^2 + 2\beta(\theta_2);$$
$$\Delta(\theta_2) = \gamma_1\gamma_2 + m_2 L_1^2 I_{S2} + [\alpha(\theta_2)]^2 > 0.$$

To establish the closed-loop control system, all the nonlinear terms in Eq. (4.1) are cancelled first, shown as Eq. (4.3).

$$\begin{cases} \tau_1 = (a_{(0)} + b_{(0)})\ddot{\theta}_1 + a_{(0)}\ddot{\theta}_2 \\ \tau_2 = a_{(0)}\ddot{\theta}_1 + a_{(0)}\ddot{\theta}_2 \end{cases} \tag{4.3}$$

Then, we obtain

$$\begin{cases} \ddot{\theta}_1 = \frac{1}{b_{(0)}}\tau_1 - \frac{1}{b_{(0)}}\tau_2 \\ \ddot{\theta}_2 = -\frac{1}{b_{(0)}}\tau_1 + (\frac{1}{a_{(0)}} + \frac{1}{b_{(0)}})\tau_2 \end{cases} \tag{4.4}$$

According to the same derivation process in Chap. 2, the closed-loop control law of manipulator_0 can be deduced as

$$\begin{cases} \tau_1 = [a_{(0)} + b_{(0)}]\ddot{\theta}_{1R} + a_{(0)}\ddot{\theta}_{2R} \\ \quad -g_{11}[(a_{(0)} + b_{(0)})[\dot{\theta}_1 - \dot{\theta}_{1R}] + a_{(0)}[\dot{\theta}_2 - \dot{\theta}_{2R}]] \\ \quad -g_{12}[(a_{(0)} + b_{(0)})[\theta_1 - \theta_{1R}] + a_{(0)}[\theta_2 - \theta_{2R}]] \\ \tau_2 = a_{(0)}\ddot{\theta}_{1R} + a_{(0)}\ddot{\theta}_{2R} \\ \quad -g_{21}[a_{(0)}[\dot{\theta}_1 - \dot{\theta}_{1R}] + a_{(0)}[\dot{\theta}_2 - \dot{\theta}_{2R}]] \\ \quad -g_{22}[a_{(0)}[\theta_1 - \theta_{1R}] + a_{(0)}[\theta_2 - \theta_{2R}]] \end{cases} \tag{4.5}$$

where, the feedback factors are: $g_{11} = g_{21} = \frac{2\sqrt{1+\sqrt{3}}}{T_p}$ and $g_{12} = g_{22} = \frac{2\sqrt{3}}{T_p^2}$.

According to the results in Chap. 2, the dynamic equation of manipulator_1 without payload is expressed as

$$\begin{cases} \tau_1 = (a_{(1)} + b_{(1)})\ddot{\theta}_1 + a_{(1)}\ddot{\theta}_2 \\ \tau_2 = a_{(1)}\ddot{\theta}_1 + a_{(1)}\ddot{\theta}_2 \end{cases} \tag{4.6}$$

where,

$$a_{(1)} = I_{S2} + I_{SR} + m_2 L_{BS2}^2 + m_{SR} L_{BS2}^2 + m_{cw} L_{cw}^2$$
$$b_{(1)} = m_1 L_{AS1}^2 + I_{S1} + m_{2+SR+cw} L_1^2$$

The structure model of manipulator_1 for the simulation is given as

$$\begin{cases} \ddot{\theta}_1 = \frac{\gamma_2}{\Delta(\theta_2)}\tau_1 - \frac{[\gamma_2 + \beta(\theta_2)]}{\Delta(\theta_2)}\tau_2 + \frac{\gamma_2\alpha(\theta_2)}{\Delta(\theta_2)}\dot{\theta}_1\dot{\theta}_2 \\ \quad + \frac{\alpha(\theta_2)[\gamma_2 + \beta(\theta_2)]}{\Delta(\theta_2)}\dot{\theta}_1^2 + \frac{\gamma_2\alpha(\theta_2)}{\Delta(\theta_2)}(\dot{\theta}_1 + \dot{\theta}_2)\dot{\theta}_2 \\ \ddot{\theta}_2 = -\frac{[\gamma_2 + \beta(\theta_2)]}{\Delta(\theta_2)}\tau_1 + \frac{\psi(\theta_2)}{\Delta(\theta_2)}\tau_2 - \frac{\alpha(\theta_2)[\gamma_2 + \beta(\theta_2)]}{\Delta(\theta_2)}\dot{\theta}_1\dot{\theta}_2 \\ \quad - \frac{\alpha(\theta_2)\psi(\theta_2)}{\Delta(\theta_2)}\dot{\theta}_1^2 - \frac{\alpha(\theta_2)[\gamma_2 + \beta(\theta_2)]}{\Delta(\theta_2)}(\dot{\theta}_1 + \dot{\theta}_2)\dot{\theta}_2 \end{cases} \tag{4.7}$$

where,

$$\alpha(\theta_2) = m_{2+SR+cw} L_1 L_{BS2r} \sin(\theta_2);$$
$$\beta(\theta_2) = m_{2+SR+cw} L_1 L_{BS2r} \cos(\theta_2);$$
$$\gamma_1 = I_{S1} + m_1 L_{AS1}^2;$$
$$\gamma_2 = I_{S2+SR+cw} + m_{2+SR+cw} L_{BS2r}^2;$$
$$\psi(\theta_2) = \gamma_1 + \gamma_2 + m_{2+SR+cw} L_1^2 + 2\beta(\theta_2);$$
$$\Delta(\theta_2) = \gamma_1\gamma_2 + m_{2+SR+cw} L_1^2 I_{S2} + [\alpha(\theta_2)]^2 > 0.$$

To establish the closed-loop system of manipulator_1, the Eq. (4.6) is reformed as

$$\begin{cases} \ddot{\theta}_1 = \frac{1}{b_{(1)}}\tau_1 - \frac{1}{b_{(1)}}\tau_2 \\ \ddot{\theta}_2 = -\frac{1}{b_{(1)}}\tau_1 + (\frac{1}{a_{(1)}} + \frac{1}{b_{(1)}})\tau_2 \end{cases} \tag{4.8}$$

According to the results in Chap. 2, the closed-loop control law of manipulator_1 can be written as

$$\begin{cases} \tau_1 = (a_{(1)} + b_{(1)})\ddot{\theta}_{1R} + a_{(1)}\ddot{\theta}_{2R} - g_{11}[(a_{(1)} + b_{(1)})[\dot{\theta}_1 - \dot{\theta}_{1R}] + a_{(1)}[\dot{\theta}_2 - \dot{\theta}_{2R}]] \\ \quad -g_{12}[(a_{(1)} + b_{(1)})[\theta_1 - \theta_{1R}] + a_{(1)}[\theta_2 - \theta_{2R}]] \\ \tau_2 = a_{(1)}\ddot{\theta}_{1R} + a_{(1)}\ddot{\theta}_{2R} - g_{21}[a_{(1)}[\dot{\theta}_1 - \dot{\theta}_{1R}] + a_{(1)}[\dot{\theta}_2 - \dot{\theta}_{2R}]] \\ \quad -g_{22}[a_{(1)}[\theta_1 - \theta_{1R}] + a_{(1)}[\theta_2 - \theta_{2R}]] \end{cases} \tag{4.9}$$

where, the feedback factors are: $g_{11} = g_{21} = \frac{2\sqrt{1+\sqrt{3}}}{T_p}$ and $g_{12} = g_{22} = \frac{2\sqrt{3}}{T_p^2}$.

According to the results in Chap. 3, the dynamic equation of manipulator_2 without payload is expressed as

$$\begin{cases} \tau_1 = a_{(2)}\ddot{\theta}_1 \\ \tau_2 = b_{(2)}\ddot{\theta}_2 \end{cases} \tag{4.10}$$

where,

$$a_{(2)} = I_{S1} + 2I_{S2} + I_{S4} + m_1 L_{AS1}^2 + (m_2 + m_3 + m_4)L_1^2 + 2\frac{(m_3 + m_4)(m_2 + m_3 + m_4)}{m_2} L_3^2$$
$$b_{(2)} = 2[I_{S2} + \frac{(m_3 + m_4)(m_2 + m_3 + m_4)}{m_2} L_3^2]$$

Thus the structure model of manipulator_2 can be described as Eqs. (4.11) and (4.12).

$$\begin{aligned}\ddot{\theta}_1 &= \frac{a_3 + a_4 - 2a_7 + 4\beta_1(\theta_2)}{\Delta(\theta_2)}\tau_1 + \frac{-a_3 + a_4 - \gamma_1(\theta_2)}{\Delta(\theta_2)}\tau_2 \\ &+ \frac{2[a_3 + a_4 - 2a_7 + 4\beta_1(\theta_2)]\,[\alpha_2(\theta_2) + 4\beta_2(\theta_2)\,]}{\Delta(\theta_2)}\dot{\theta}_1\dot{\theta}_2 \\ &- \frac{[-a_3 + a_4 - \gamma_1(\theta_2)][\alpha_2(\theta_2) + 4\beta_2(\theta_2)]}{\Delta(\theta_2)}\dot{\theta}_1^2 \\ &+ \frac{[a_3 + a_4 - 2a_7 + 4\beta_1(\theta_2)][\gamma_2(\theta_2)]}{\Delta(\theta_2)}\dot{\theta}_2^2 - \frac{4[-a_3 + a_4 - \gamma_1(\theta_2)][\beta_2(\theta_2)]}{\Delta(\theta_2)}\dot{\theta}_2^2\end{aligned} \tag{4.11}$$

$$\begin{aligned}\ddot{\theta}_2 &= \frac{-a_3 + a_4 - \gamma_1(\theta_2)}{\Delta(\theta_2)}\tau_1 + \frac{a_0 + 2\alpha_1(\theta_2) - 4\beta_1(\theta_2)}{\Delta(\theta_2)}\tau_2 \\ &+ \frac{2[-a_3 + a_4 - \gamma_1(\theta_2)][\alpha_2(\theta_2) + 4\beta_2(\theta_2)\,]}{\Delta(\theta_2)}\dot{\theta}_1\dot{\theta}_2 \\ &- \frac{[a_0 + 2\alpha_1(\theta_2) - 4\beta_1(\theta_2)][\alpha_2(\theta_2) + 4\beta_2(\theta_2)]}{\Delta(\theta_2)}\dot{\theta}_1^2 \\ &+ \frac{[-a_3 + a_4 - \gamma_1(\theta_2)]\,[\gamma_2(\theta_2)]}{\Delta(\theta_2)}\dot{\theta}_2^2 - \frac{4[a_0 + 2\alpha_1(\theta_2) - 4\beta_1(\theta_2)][\beta_2(\theta_2)]}{\Delta(\theta_2)}\dot{\theta}_2^2\end{aligned} \tag{4.12}$$

where,

$$\begin{aligned}a_0 &= [m_1L_{AS1}^2 + I_{S1} + M_2L_1^2 + M_2L_{BS2r}^2 + I_{2+p} + m_3L_1^2 + m_3L_3^2 \\ &\quad + m_3L_{CS3}^2 + I_{S3} + I_{S4} + 4m_4L_3^2 + m_4L_1^2 + 2m_3L_3L_{CS3}]; \\ a_1 &= [m_3L_3L_{CS3} + m_4L_3^2]; \\ a_2 &= [M_2L_1L_{BS2r} - m_3L_1L_3]; \\ a_3 &= [M_2L_{BS2r}^2 + I_{2+p} + m_3L_3^2]; \\ a_4 &= [m_3L_{CS3}^2 + I_3]; \\ a_5 &= m_3L_1L_3; \\ a_6 &= m_4L_1L_3; \\ a_7 &= m_3L_3L_{CS3}; \\ \alpha_1(\theta_2) &= [a_2 - a_5 - a_6]\cos(\theta_2); \\ \alpha_2(\theta_2) &= [a_2 - a_5 - a_6]\sin(\theta_2); \\ \beta_1(\theta_2) &= a_1\sin^2(\theta_2); \\ \beta_2(\theta_2) &= a_1\sin(\theta_2)\cos(\theta_2);\end{aligned}$$

$$\gamma_1(\theta_2) = [a_2 + a_5]\cos(\theta_2);$$
$$\gamma_2(\theta_2) = [a_2 + a_5]\sin(\theta_2).$$

As the same, to establish the closed-loop control of manipulator_2, the Eq. (4.10) is reformed as

$$\begin{cases} \ddot{\theta}_1 = \frac{1}{a_{(2)}}\tau_1 \\ \ddot{\theta}_2 = \frac{1}{b_{(2)}}\tau_2 \end{cases} \tag{4.13}$$

According to the results in Chap. 3, the closed-loop control law of manipulator_2 can be written as

$$\begin{cases} \tau_1 = a_{(2)}\ddot{\theta}_{1R} - g_{11}a_{(2)}[\dot{\theta}_1 - \dot{\theta}_{1R}] - g_{12}a_{(2)}[\theta_1 - \theta_{1R}] \\ \tau_2 = b_{(2)}\ddot{\theta}_{2R} - g_{21}b_{(2)}[\dot{\theta}_2 - \dot{\theta}_{2R}] - g_{22}b_{(2)}[\theta_2 - \theta_{2R}] \end{cases} \tag{4.14}$$

where, the feedback factors are: $g_{11} = g_{21} = \frac{2\sqrt{1+\sqrt{3}}}{T_p}$ and $g_{12} = g_{22} = \frac{2\sqrt{3}}{T_p^2}$.

The manipulator_3 has the same structure as manipulator_0 (Fig. 4.4), however, it is linearized and decoupled by feedback linearization which is shown in Fig. 4.5.

The dynamic model of manipulator_3 can be expressed as

$$\tau = A(\theta)\ddot{\theta} + C(\theta, \dot{\theta})\dot{\theta} \tag{4.15}$$

The decoupling law is written as

$$\tau = \hat{A}(\theta)v + \hat{C}(\theta, \dot{\theta})\dot{\theta} \tag{4.16}$$

where, $\hat{A}(\theta)$ is referred to as the inertia matrix that be formed by nominal parameters; $\hat{C}(\theta, \dot{\theta})$ is referred to as Coriolis and centrifugal effects that be formed by nominal parameters.

If $\hat{A}(\theta) = A(\theta)$ and $\hat{C}(\theta, \dot{\theta}) = C(\theta, \dot{\theta})$, then we obtain

$$A(\theta)v = A(\theta)\ddot{\theta} \tag{4.17}$$

In addition, the inertia matrix $A(\theta)$ is reversible, thus

$$v(t) = \ddot{\theta}(t) \tag{4.18}$$

Substituting Eqs. (4.15) and (4.18) into Eq. (4.1), then the equation of the torque can be rewritten as

$$\begin{cases} \tau_1 = [a_{(0)} + b_{(0)} + 2m_2L_1L_{BS2}\cos\theta_2]v_1 + [a_{(0)} + m_2L_1L_{BS2}\cos\theta_2]v_2 \\ \qquad + [-m_2L_1L_{BS2}\sin\theta_2]\dot{\theta}_2^2 + [-2m_2L_1L_{BS2}\sin\theta_2]\dot{\theta}_1\dot{\theta}_2 \\ \tau_2 = [a_{(0)} + m_2L_1L_{BS2}\cos\theta_2]v_1 + a_{(0)}v_2 + [m_2L_1L_{BS2}\sin\theta_2]\dot{\theta}_1^2 \end{cases} \tag{4.19}$$

Replacing τ_1 and τ_2 of the Eq. (4.1) by Eq. (4.19), finally, we can obtain

$$\begin{bmatrix} \ddot{\theta}_1(t) \\ \ddot{\theta}_2(t) \end{bmatrix} = \begin{bmatrix} v_1(t) \\ v_2(t) \end{bmatrix} \tag{4.20}$$

Now, as shown in Fig. 4.5, the nonlinear manipulator which is described by Eq. (4.2) becomes linearized and decoupled.

Based on this linear and decoupled model, the closed-loop control system can be furthermore established. Because that this model is similar to manipulator_2, according to the same derivation process, the closed-loop control law and the feedback factors can be obtained

$$\begin{cases} v_1(t) = \ddot{\theta}_{1R}(t) - g_{11}[\dot{\theta}_1(t) - \dot{\theta}_{1R}(t)] - g_{12}[\theta_1(t) - \theta_{1R}(t)] \\ v_2(t) = \ddot{\theta}_{2R}(t) - g_{21}[\dot{\theta}_2(t) - \dot{\theta}_{2R}(t)] - g_{22}[\theta_2(t) - \theta_{2R}(t)] \end{cases} \tag{4.21}$$

where, the feedback factors are: $g_{11} = g_{21} = \frac{2\sqrt{1+\sqrt{3}}}{T_p}$ and $g_{12} = g_{22} = \frac{2\sqrt{3}}{T_p^2}$.

Through the comparisons among the closed-loop control law of these four manipulators, it is found that, the manipulator_0 is similar to manipulator_1, and the manipulator_2 is similar to manipulator_3. However, in the aspect of the structure model, the manipulator_1 and manipulator_2 are much simpler than manipulator_0 and manipulator_3.

Next, in order to compare the robustness of these manipulators, the error tolerance capability analysis is introduced. In addition, to make meaningful comparisons between different manipulators, it is useful to quantify the system capability in terms of a representative numerical index. This also allows one to optimize the design with respect to the physical parameters such as the mass distribution, actuator location or link dimensions. Hence, here in order to quantitatively describe the tolerance capability against the parametric variation, two kinds of indices referred as angular error and position error are identified.

The angular error is defined as the ratio of the absolute angular variation to the desired overall angular displacement $\Delta\theta_i$. It is calculated as

$$\varepsilon_{\theta i} = \left|\frac{\delta\theta_i}{\Delta\theta_i}\right| \times 100\% \quad (i = 1, 2) \tag{4.22}$$

where, $\Delta\theta_i = \theta_{iF} - \theta_{iI}$. θ_{iI} represents the desired initial angle of the *ith* actuator, θ_{iF} represents the desired final angle of the *ith* actuator.

According to the kinematic of the manipulators, the position vector of the end-effector of the serial manipulator in the Cartesian coordinate that attached on the first joint can be written as

$$P = \begin{bmatrix} \cos\theta_1 & \cos(\theta_1 + \theta_2) \\ \sin\theta_1 & \sin(\theta_1 + \theta_2) \end{bmatrix} \begin{bmatrix} L_1 \\ L_2 \end{bmatrix} \tag{4.23}$$

where, $P = \begin{bmatrix} x_P \\ y_P \end{bmatrix}$.

The position error is defined as the norm of the position vector variation that caused by the variation of the angular displacement of the actuators. The expression is described as

$$\varepsilon_P = \sqrt{\delta x_P^2 + \delta y_P^2} \quad \text{(mm)} \tag{4.24}$$

with,

$$\begin{aligned} \delta x_P &= L_{1N}\cos\theta_{1N} + L_{2N}\cos(\theta_{1N} + \theta_{2N}) - L_{1R}\cos\theta_{1R} - L_{2R}\cos(\theta_{1R} + \theta_{2R}) \\ \delta y_P &= L_{1N}\sin\theta_{1N} + L_{2N}\sin(\theta_{1N} + \theta_{2N}) - L_{1R}\sin\theta_{1R} - L_{2R}\sin(\theta_{1R} + \theta_{2R}) \end{aligned} \tag{4.25}$$

where, δx_P, δy_P are the components of the position vector variation on the x axis and y axis of the Cartesian coordinate respectively; $L_{1N}, L_{2N}, \theta_{1N}, \theta_{2N}$ are the nominal parameters; $L_{1R}, L_{2R}, \theta_{1R}, \theta_{2R}$ are the real parameters.

4.3 Tolerance Capability Comparison Among the Manipulators

For the comparison, the main parameters are needed to be uniformed, shown as in Table 4.1. As the special condition of reverse rotations of the two main links is necessary for the dynamic decoupling of manipulator_1, the initial and final angular positions of the manipulators are given as in Table 4.2.

The desired trajectories of the two links are produced by Eqs. (2.29) and (2.30). With these simulation conditions, two kinds of comparison methods are designed,

Table 4.1 The parametric values of the open-loop control system

Mass (kg)		Length (m)		Moment of inertia (kg m^2)	
m_1	13.193	L_1/L_{AS1}	0.8/0.4	I_{S1}	0.77754
m_2	8.477	L_2/L_{BS2}	0.5/0.25	I_{S2}	0.20744

Table 4.2 The initial and final values of the desired trajectories

Angle (°)		Velocity (m/s)		Acceleration (m/s^2)	
θ_{1I}	60	$\dot{\theta}_{1I}$	0	$\ddot{\theta}_{1I}$	0
θ_{1F}	0	$\dot{\theta}_{1F}$	0	$\ddot{\theta}_{1F}$	0
θ_{2I}	20	$\dot{\theta}_{2I}$	0	$\ddot{\theta}_{2I}$	0
θ_{2F}	80	$\dot{\theta}_{2F}$	0	$\ddot{\theta}_{2F}$	0

they are implemented by the introduction of fixed and random parametric errors respectively. All the simulations are taken under the close-loop systems.

4.3.1 *Tolerance Capability Comparison by Introducing the Fixed Parametric Error*

In this section, to provide insight to the tolerance capability, a set of simulations are designed for the main independent variables. In each simulation, the error is added to one and only one parameter, the value of error is assumed to vary from −20 to 20% in step of 5%. Here, δm_1, δm_2, δI_{S1}, δI_{S2}, δL_{AS1} and δL_{BS2} are noted as the diviation of the variables m_1, m_2, I_{S1}, I_{S2} and L_{AS1} respectivily.

During these simulations, the angular error $\varepsilon_{\theta i}$ and the position error ε_P are calculated according to the Eqs. (4.22) and (4.24).

The simulations with 20% parametric errors of the variables are shown in Fig. 4.6 to compare the tracking accuracy among the four manipulators during the whole process. Each figure shows the influence on the angular accuracy of one link by introducing one kind of variable error for the four manipulators.

Figure 4.6a1 shows the curves of absolut deviation of θ_1 that influenced by δm_1 during the whole process. Obviously, for the coupled manipulator_0, there are three peaks of wave, for all the three dynamic decoupled models (manipulator_1, manipulator_2, manipulator_3), there are two peaks of wave. Moreover, they are much lower than the peaks of wave in the coupled manipulator_0. It should be noted that, during the coupled manipulators, the angular error curves of manipulator_1 and manipulator_2 are similar and both of theirs peaks of wave are lower than the ones of manipulator_3 which is dynamic decoupled by the control method. However, with the increase of operation time, for all the manipulators, the difference between the desired trajectory and the real trajectory became smaller and smaller, and then became a constant.

From this aspect, the rates of convergence of the decoupled manipulators are similar and faster than the coupled manipulator_0. The similar phenomenon also exists in Fig. 4.6a2, a3. The robust of the decoupled manipulators for tracking the desired trajectory also displayed in Fig. 4.6a5, a6, especially in Fig. 4.6a4. It shows that, during the whole process, the parametric error of variable I_{S2} has no influence on the tracking accuracy of link 1.

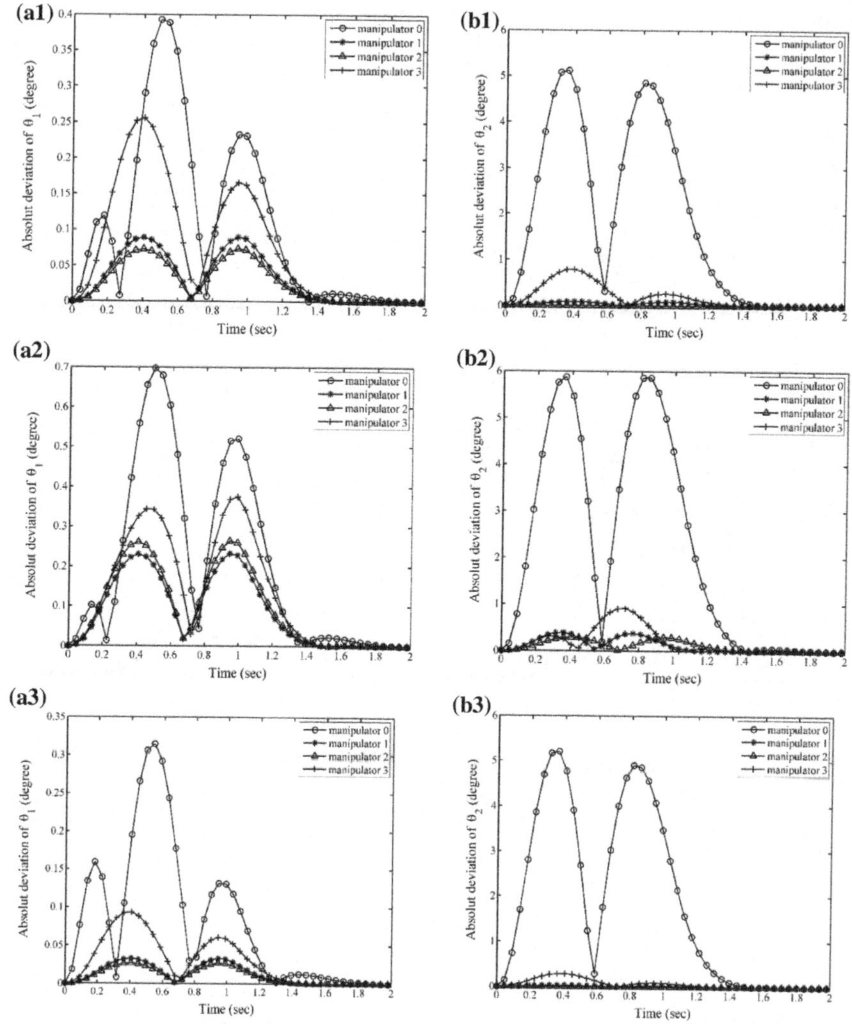

Fig. 4.6 a1–b6 The absolut deviation of the two links that influenced by introducing 20% parametric errors of the variables during the whole process of the four manipulators. **a1–a6** Show the angular errors of link 1 that influenced by δm_1, δm_2, δI_{S1}, δI_{S2}, δL_{AS1}, δL_{BS2}, respectively. **b1–b6** Show the angular errors of link 2 that influenced by δm_1, δm_2, δI_{S1}, δI_{S2}, δL_{AS1}, δL_{BS2} respectively

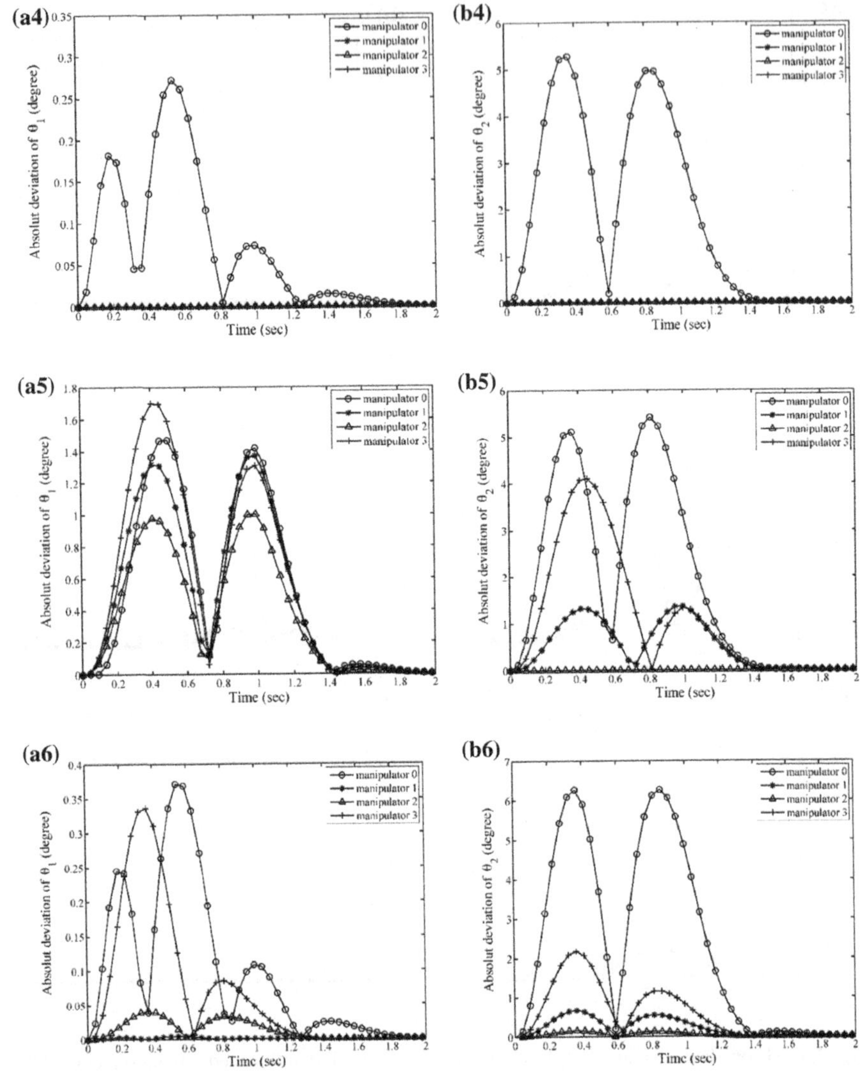

Fig. 4.6 (continued)

Figure 4.6b1–b6 shows the curves of angular error of link 2 that influenced by δm_1, δm_2, δI_{S1}, δI_{S2}, δL_{AS1}, δL_{BS2} during the whole process respectively. All of them strongly proved that the decoupled manipulators can track the desired trajectory more precise and less sensitive to the variable errors during the whole process.

In addition, it also can be found that, the tracking accuracies of manipulator_1 and manipulator_2 are higher than the ones of manipulator_3.

Moreover, according to the final tracking results, the higher robust performance of the decoupled manipulator has also been proved (Tables 4.3, 4.4, 4.5 and 4.6).

Table 4.3 shows the influence on the positioning accuracy of manipulator_0. As mentioned above, in order to establish the closed-loop control, all the nonlinear terms in the dynamic equations are cancelled. This causes the discordance between the control model and the real manipulator.

The result is that, even not any parametric error is introduced, the positioning error still exists. For all kinds of variables, along with the increase of the parametric error, both the angular error and the position error become larger.

Tables 4.4, 4.5 and 4.6 show the influences on the positioning accuracies of manipulator_1, manipulator_2 and manipulator_3 respectively. They are all linearized and decoupled models. Through the comparison between the linear model and the nonlinear model, the biggest difference is that, there is no positioning error in the simulations of the linear models with the nominal parametric values.

In addition, compared with the values of both angular error and position error which are caused by introducing δm_1, δm_2, δI_{S1}, and δI_{S2} respectively in the coupled manipulator_0, the values of the corresponding ones in the decoupled manipulators are an order of magnitude. Especially, the positioning errors that caused by δI_{S2} in the manipulator_1, manipulator_2 and manipulator_3 are all zero, that is, the inertia deviation of link 2 has no influence on the positioning error in the decoupled manipulators. Moreover, it can be found that, the angular accuracy of link 2 can't be affected by the variable deviations of link 1 (δm_1, δI_{S1}, δL_{AS1}), this is also confirmed in Fig. 4.6a4, b4 above.

Compared with other kinds of variables, the parametric errors of the length variables lead to the biggest influences on the positioning accuracy. It seems that the control system here cannot compensate the position error that caused by the variation of length variable. Besides, the effects of the mass variables are weaker. And the third impact factor for positioning accuracy is the inertia variable.

4.3.2 *Tolerance Capability Comparison by Introducing the Random Parametric Error*

In last section, the influencing degrees on the angular error and position error that caused by the six variables are analyzed respectively. In order to closer the practical situation, the variations of all the variables are added to the nominal models at the same time. As the control method cannot compensate the positon errors that caused by the length variables, only the mass diviations (δm_1, δm_2) and inertia diviations (δI_{S1}, δI_{S2}) are investigated in this section.

Thus, the positioning accuracy of the manipulator is influenced by the parametric errors of all the variables together.

There are totally eleven simulations for each manipulator. One of them is implemented with nominal parametric values. The rest ten simulations are

Table 4.3 The influence on the positioning accuracy of manipulator_0 by introducing fixed errors

		−20%	−15%	−10%	−5%	0	5%	10%	15%	20%
δm_1	$\varepsilon_{\theta 1}$	0.00024	0.00023	0.00023	0.00023	0.00024	0.00025	0.00026	0.00028	0.00029
	$\varepsilon_{\theta 2}$	0.00074	0.00082	0.00089	0.00095	0.00101	0.00106	0.00110	0.00114	0.00118
	ε_p	0.00591	0.00622	0.00655	0.00687	0.00721	0.00754	0.00788	0.00822	0.00856
δm_2	$\varepsilon_{\theta 1}$	0.00031	0.00027	0.00024	0.00024	0.00024	0.00024	0.00023	0.00022	0.00018
	$\varepsilon_{\theta 2}$	0.00111	0.00051	0.00004	0.00056	0.00101	0.00137	0.00163	0.00178	0.00179
	ε_p	0.00453	0.00240	0.00280	0.00498	0.00721	0.00906	0.01036	0.01095	0.01074
δI_{S1}	$\varepsilon_{\theta 1}$	0.00023	0.00023	0.00023	0.00024	0.00024	0.00024	0.00025	0.00025	0.00025
	$\varepsilon_{\theta 2}$	0.00092	0.00094	0.00097	0.00099	0.00101	0.00103	0.00105	0.00106	0.00108
	ε_p	0.00672	0.00684	0.00696	0.00708	0.00721	0.00733	0.00745	0.00758	0.00770
δI_{S2}	$\varepsilon_{\theta 1}$	0.00030	0.00029	0.00027	0.00026	0.00024	0.00022	0.00020	0.00019	0.00017
	$\varepsilon_{\theta 2}$	0.00047	0.00062	0.00076	0.00089	0.00101	0.00111	0.00120	0.00127	0.00133
	ε_p	0.00521	0.00577	0.00630	0.00678	0.00721	0.00758	0.00788	0.00812	0.00829
δL_{AS1}	$\varepsilon_{\theta 1}$	0.00153	0.00089	0.00046	0.00025	0.00024	0.00039	0.00062	0.00091	0.00129
	$\varepsilon_{\theta 2}$	0.00221	0.00111	0.00019	0.00052	0.00101	0.00128	0.00138	0.00136	0.00129
	ε_p	159.996	119.998	80.001	40.003	0.00721	39.991	79.989	119.988	159.986
δL_{BS2}	$\varepsilon_{\theta 1}$	0.00030	0.00031	0.00031	0.00028	0.00024	0.00016	0.00005	0.00009	0.00028
	$\varepsilon_{\theta 2}$	0.00195	0.00120	0.00041	0.00036	0.00101	0.00140	0.00137	0.00073	0.00073
	ε_p	99.997	74.997	49.997	24.997	0.00721	25.0013	50.000	74.999	99.997

Table 4.4 The influence on the positioning accuracy of manipulator_1 by introducing fixed errors

		−20%	−15%	−10%	−5%	0	5%	10%	15%	20%
δm_1	$\varepsilon_{\theta 1}$	0.00005	0.00004	0.00002	0.00001	0	0.00001	0.00003	0.00005	0.00007
	$\varepsilon_{\theta 2}$	0.00005	0.00004	0.00002	0.00001	0	0.00001	0.00003	0.00005	0.00007
	ε_p	0.00044	0.00034	0.00024	0.00012	0	0.00013	0.00028	0.00044	0.00061
δm_2	$\varepsilon_{\theta 1}$	0.00008	0.00007	0.00006	0.00003	0	0.00004	0.00009	0.00015	0.00022
	$\varepsilon_{\theta 2}$	0.00002	0.00001	0.00000	0.000001	0	0.00000	0.00001	0.00002	0.00003
	ε_p	0.00098	0.00087	0.00067	0.00038	0	0.00046	0.00100	0.00161	0.00228
δI_{S1}	$\varepsilon_{\theta 1}$	0.00002	0.00001	0.00001	0.00000	0	0.00000	0.00001	0.00001	0.00002
	$\varepsilon_{\theta 2}$	0.00002	0.00001	0.00001	0.000006	0	0.00000	0.00001	0.00001	0.00002
	ε_p	0.00018	0.00014	0.00009	0.00004	0	0.00005	0.00010	0.00015	0.00020
δI_{S2}	$\varepsilon_{\theta 1}$	0	0	0	0	0	0	0	0	0
	$\varepsilon_{\theta 2}$	0	0	0	0	0	0	0	0	0
	ε_p	0	0	0	0	0	0	0	0	0
δL_{AS1}	$\varepsilon_{\theta 1}$	0.00130	0.00057	0.00010	0.00008	0	0.00031	0.00079	0.00136	0.00203
	$\varepsilon_{\theta 2}$	0.00130	0.00057	0.00010	0.00008	0	0.00031	0.00079	0.00136	0.00203
	ε_p	160.000	120.000	80.000	40.000	0	40.000	80.000	120.000	160.000
δL_{BS2}	$\varepsilon_{\theta 1}$	0	0	0	0	0	0	0	0	0
	$\varepsilon_{\theta 2}$	0.00016	0.00012	0.00008	0.00004	0	0.00004	0.00008	0.00012	0.00017
	ε_p	99.999	74.999	49.999	24.999	0	25.000	50.000	75.000	100.000

Table 4.5 The influence on the positioning accuracy of manipulator_2 by introducing fixed errors

		−20%	−15%	−10%	−5%	0	5%	10%	15%	20%
δm_1	$\varepsilon_{\theta 1}$	0.00004	0.00003	0.00002	0.00001	0	0.00001	0.00002	0.00004	0.00005
	$\varepsilon_{\theta 2}$	0	0	0	0	0	0	0	0	0
	ε_p	0.00047	0.00037	0.00025	0.00013	0	0.00014	0.00029	0.00045	0.00062
δm_2	$\varepsilon_{\theta 1}$	0.00008	0.00008	0.00006	0.00004	0	0.00005	0.00011	0.00018	0.00026
	$\varepsilon_{\theta 2}$	0.00008	0.00008	0.00007	0.00004	0	0.00005	0.00012	0.00020	0.00029
	ε_p	0.00073	0.00070	0.00057	0.000333	0	0.00042	0.00093	0.00151	0.00217
δI_{S1}	$\varepsilon_{\theta 1}$	0.00001	0.00001	0.00000	0.000005	0	0.000005	0.00001	0.00001	0.00002
	$\varepsilon_{\theta 2}$	0	0	0	0	0	0	0	0	0
	ε_p	0.00019	0.00014	0.00009	0.00005	0	0.00005	0.00010	0.00015	0.00021
δI_{S2}	$\varepsilon_{\theta 1}$	0	0	0	0	0	0	0	0	0
	$\varepsilon_{\theta 2}$	0	0	0	0	0	0	0	0	0
	ε_p	0	0	0	0	0	0	0	0	0
δL_{AS1}	$\varepsilon_{\theta 1}$	0.00050	0.00017	0.00003	0.000087	0	0.00021	0.00054	0.00094	0.00138
	$\varepsilon_{\theta 2}$	0	0	0	0	0	0	0	0	0
	ε_p	160.002	120.000	79.999	39.999	0	39.998	79.997	119.991	159.992
δL_{BS2}	$\varepsilon_{\theta 1}$	0	0	0	0	0	0	0	0	0
	$\varepsilon_{\theta 2}$	0.00003	0.00002	0.00001	0.00001	0	0.00001	0.00002	0.00003	0.00004
	ε_p	100.000	75.000	50.000	25.000	0	25.000	50.000	75.000	100.000

Table 4.6 The influence on the positioning accuracy of manipulator_3 by introducing fixed errors

		−20%	−15%	−10%	−5%	0	5%	10%	15%	20%
δm_1	$\varepsilon_{\theta 1}$	0.00007	0.00006	0.00004	0.00002	0	0.00003	0.00006	0.00010	0.00014
	$\varepsilon_{\theta 2}$	0.00012	0.00010	0.00007	0.00004	0	0.00004	0.00009	0.00015	0.00021
	ε_p	0.00066	0.00055	0.00040	0.00022	0	0.00025	0.00054	0.00087	0.00122
δm_2	$\varepsilon_{\theta 1}$	0.00003	0.00007	0.00008	0.00005	0	0.00008	0.00019	0.00031	0.00046
	$\varepsilon_{\theta 2}$	0.00007	0.00005	0.00004	0.00002	0	0.00004	0.00011	0.00021	0.00036
	ε_p	0.00031	0.00065	0.00075	0.00052	0	0.00076	0.00171	0.00280	0.00398
δI_{S1}	$\varepsilon_{\theta 1}$	0.00003	0.00002	0.00002	0.00001	0	0.00001	0.00002	0.00003	0.00004
	$\varepsilon_{\theta 2}$	0.00005	0.00004	0.00003	0.00001	0	0.00001	0.00003	0.00005	0.00007
	ε_p	0.00031	0.00024	0.00016	0.00008	0	0.00009	0.00018	0.00028	0.00039
δI_{S2}	$\varepsilon_{\theta 1}$	0	0	0	0	0	0	0	0	0
	$\varepsilon_{\theta 2}$	0	0	0	0	0	0	0	0	0
	ε_p	0	0	0	0	0	0	0	0	0
δL_{AS1}	$\varepsilon_{\theta 1}$	0.00114	0.00049	0.00008	0.000085	0	0.00030	0.00077	0.00135	0.00206
	$\varepsilon_{\theta 2}$	0.00134	0.00059	0.00010	0.00009	0	0.00035	0.00090	0.00158	0.00242
	ε_p	159.998	119.999	79.999	40.000	0	40.0002	80.0006	120.001	160.001
δL_{BS2}	$\varepsilon_{\theta 1}$	0.00001	0.00000	0.00000	0.00000	0	0.00000	0.00000	0.00001	0.00001
	$\varepsilon_{\theta 2}$	0.00029	0.00008	0.00003	0.00005	0	0.00008	0.00015	0.00011	0.00017
	ε_p	99.999	74.999	49.999	24.999	0	24.999	49.999	74.999	99.999

implemented by adding the parametric errors of the four variables in the same time. In order to let the parametric values closer to the practical situation, ten sets of random values are created in MATLAB first, shown as

$$Ran = \begin{bmatrix} 0,814724 & 0,157613 & 0,655741 & 0,706046 \\ 0,905792 & 0,970593 & 0,035712 & 0,031833 \\ 0,126987 & 0,957167 & 0,849129 & 0,276923 \\ 0,913376 & 0,485376 & 0,933993 & 0,046171 \\ 0,632359 & 0,80028 & 0,678735 & 0,097132 \\ 0,09754 & 0,141886 & 0,75774 & 0,823458 \\ 0,278498 & 0,421761 & 0,743132 & 0,694829 \\ 0,546882 & 0,915736 & 0,392227 & 0,317099 \\ 0,957507 & 0,792207 & 0,655478 & 0,950222 \\ 0,964889 & 0,959492 & 0,171187 & 0,034446 \end{bmatrix} \tag{4.26}$$

In order to make sure that the variation range of the variable is still from −20 to 20%. The Eqs. (4.27)–(4.30) are used to calculate the real parametric values for all the simulations.

$$m_{1r} = m_{1n} \times (0.8 + 0.4 \times Ran(i, 1)) \tag{4.27}$$

$$m_{2r} = m_{2n} \times (0.8 + 0.4 \times Ran(i, 2)) \tag{4.28}$$

$$I_{S1r} = I_{S1n} \times (0.8 + 0.4 \times Ran(i, 3)) \tag{4.29}$$

$$I_{S2r} = I_{S2n} \times (0.8 + 0.4 \times Ran(i, 4)) \tag{4.30}$$

where, m_{1r}, m_{2r}, I_{S1r} and I_{S2r} represent the real parametric values; m_{1n}, m_{2n}, I_{S1n} and I_{S2n} represent the nominal parametric values.

Finally, the real parametric values for all the simulations can obtained, shown in Table 4.7. In this table, Sim_i means the *i*th simulation. It should be note that, all the parametric values are nominal value in the first simulation (Sim_1). In the rest simulations (from Sim_2 to Sim_11), the parametric values are all added with errors.

Based on these simulations, the total tolerance capability of all the models can be compared. Figure 4.7 shows the angular errors of the link 1. It shows that, in Sim_1, all the angular errors of link 1 of the decoupled manipulators (manipulator_1, manipulator_2 and manipulator_3) are zero. On the contrary, for the coupled manipulator_0, there is the angular error. As discussed in the last section, this is because the 'mismatch' between the control model and the actual model in coupled manipulator_0. This phenomenon also exists for the angular error of link 2 and the positioning error of the end-effector, shown in Figs. 4.8 and 4.9.

When the parametric errors are added, the angular errors appear. It shows that the angular errors of link 1 of manipulator_1 and manipulator_2 are close and smaller than the ones of manipulator_0 and manipulator_3. Meanwhile the range of variation of the angular errors of link 1 in manipulator_1 and manipulator_2 are also smaller than the ones in manipulator_0 and manipulator_3.

The angular errors of link 2 that caused by the parametric errors of all the variables together are shown in Fig. 4.8. Obviously, the angular error of the

Table 4.7 The real parametric values for the ten simulations

	m_{1r}	m_{2r}	I_{S1r}	I_{S2r}
Sim_1	13.193	8.477	0.77754	0.20744
Sim_2	14.85386	7.316034	0.825978	0.224537
Sim_3	15.33445	10.07269	0.633139	0.168593
Sim_4	11.22453	10.02716	0.886125	0.18893
Sim_5	15.37447	8.427412	0.912519	0.169783
Sim_6	13.89149	9.495191	0.833129	0.174012
Sim_7	11.06914	7.262708	0.857701	0.234279
Sim_8	12.02409	8.211708	0.853158	0.223606
Sim_9	13.4404	9.886676	0.744021	0.192264
Sim_10	15.60736	9.467817	0.825896	0.244798
Sim_11	15.64631	10.03505	0.675274	0.16881

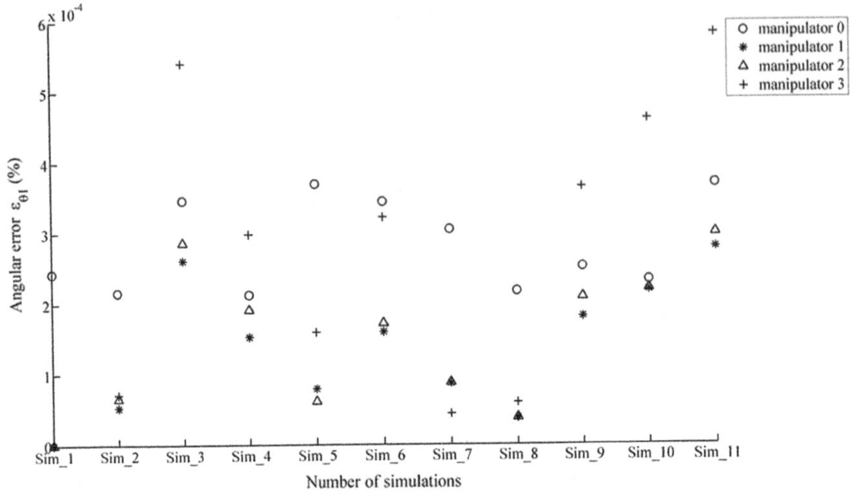

Fig. 4.7 The angular error of link 1 with the parametric errors of all the variables

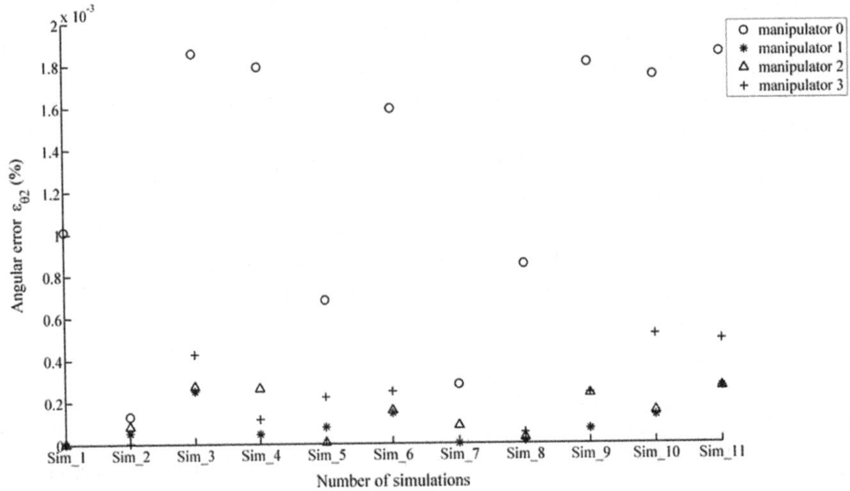

Fig. 4.8 The angular error of link 2 with the parametric errors of all the variables

coupled manipulator_0 is much higher than the ones of the decoupled manipulators. Among the three coupled manipulator, it is difficult to say which is better. This is difference with the phenomenon in Fig. 4.7.

In Fig. 4.9, the position errors of the end-effector of all the models are represented. A clear tendency appears that the position error of the end-effector in the coupled manipulator_0 is the highest. In addition, the position errors of the end-effectors in manipulator_1 and manipulator_2 are close and they are smaller

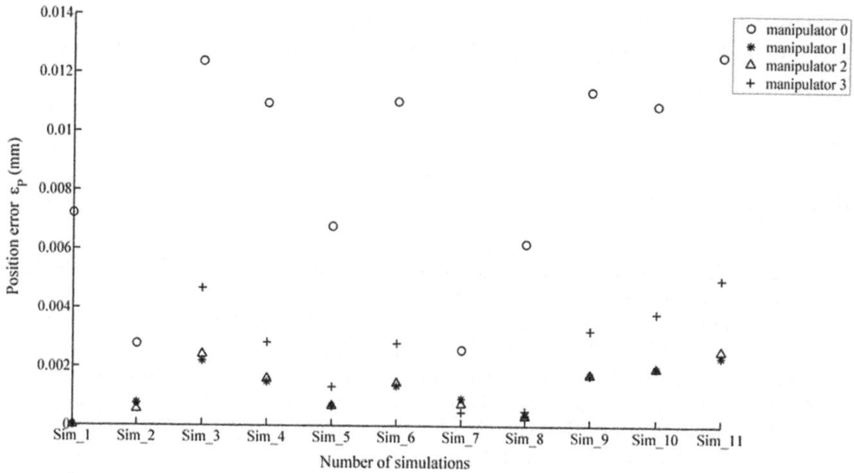

Fig. 4.9 The position error of the end-effector with the parametric errors of all the variables

than the ones in manipulator_3. In other words, the tolerance capability of manipulator_1 and manipulator_2 which are decoupled by the mechatronic methods that proposed in this work are similar. And these two manipulators are more robust than manipulator_3 which is decoupled by the control method. Of course, all these three decoupled manipulators are much more robust than the coupled manipulator.

4.4 Summary

In this chapter, the tolerance capabilities of four models are analyzed. Two kinds of indices are proposed to quantify the positioning accuracy of the manipulator. They are angular error of the actuators and the position error of the end-effector.

First, in order to analysis the influencing degree of each variable of the manipulator on the positioning accuracy, the fixed parametric errors are introduced. According to the quantitative analysis, it shows that the positioning accuracy is more sensitive with the variation of the length variables. The second influencing factors are the mass parameters. And the influences of the inertia parameters on the positioning accuracy are the lowest.

In addition, it can be also found that, during the whole process, the tracking trajectories of the decoupled manipulators, especially in manipulator_1 and manipulator_2, are more precise and less sensitive to the variable errors.

Then, in order to obtain the models closer to the practical situation, the random parametric errors are introduced. Furthermore, the parametric errors of all the variables are added at the same time during one simulation. According to the results, the advantages of the coupled manipulators (manipulator_1 and manipulator_2) appear. The tolerance capabilities of these two manipulators are higher than the ones of manipulator_3 which is decoupled by feedback linearization and the coupled manipulator.

References

1. Gosselin CM (1990) Dexterity indices for planar and spatial robotic manipulators. In: IEEE international conference on robotics and automation 1990. pp 650–655
2. Asada H (1983) A geometrical representation of manipulator dynamics and its application to arm design. J Dyn Syst Meas Control 105(3):131–142
3. Yoshikawa T (1985) Dynamic manipulability of robot manipulators. In: IEEE international conference on robotics and automation 1985, vol. 2. pp. 1033–1038
4. Ma O, Angeles J (1990) The concept of dynamic isotropy and its applications to inverse kinematics and trajectory planning. pp 481–486
5. Ben-Gharbia KM, Maciejewski AA, Roberts RG (2015) Modifying the kinematic structure of an anthropomorphic arm to improve fault tolerance. In: IEEE international conference on robotics and automation (ICRA) 2015, pp 1455–1460
6. Cheng S, Dhillon BS (2011) Reliability and availability analysis of a robot-safety system. J Qual Maint Eng 17(2):203–232
7. Dixon WE, Walker ID, Dawson DM, Hartranft JP (2000) Fault detection for robot manipulators with parametric uncertainty: a prediction-error-based approach. IEEE Trans Robot Autom 16(6):689–699
8. Ben-Gharbia KM, Roberts RG, Maciejewski AA (2011) Examples of planar robot kinematic designs from optimally fault-tolerant Jacobians. In: IEEE international conference on robotics and automation 2011, pp 4710–4715
9. Veitschegger W, Wu C-H (1986) Robot accuracy analysis based on kinematics. IEEE J Robot Autom 2(3):171–179
10. Wu CH (1983) The kinematic error model for the design of robot manipulator. In: American control conference 1983, pp 497–502
11. Veitschegger WK, Wu C-H (1988) Robot calibration and compensation. IEEE J Robot Autom 4(6):643–656
12. Waldron KJ, Kumar A (1979) Development of a theory of errors for manipulators. In: Proceedings of the fifth world congress on the theory of machines and mechanisms 1979, pp 821–826
13. Benhabib B, Fenton R, Goldenberg A (1987) Computer-aided joint error analysis of robots. IEEE J Robot Autom 3(4):317–322
14. Vaishnav RN, Magrab EB (1987) A general procedure to evaluate robot positioning errors. Int J Robot Res 6(1):59–74
15. Ferreira PM, Liu CR (1986) An analytical quadratic model for the geometric error of a machine tool. J Manuf Syst 5(1):51–63
16. Caro S, Bennis F, Wenger P (2005) Tolerance synthesis of mechanisms: a robust design approach. J Mech Des 127(1):86–94

Chapter 5
Dynamics Decoupling Control of Parallel Manipulator

Jun-Wei Han, Wei Wei and Zhi-Dong Yang

Abstract Due to the eccentric load of parallel manipulator, dynamic coupling occurs between the various degrees of freedom. As a typical parallel mechanism, the dynamic model of redundant shaking table is built. Coupling force observaction based on coupling model is introduced to the DoFs control structure. The coupling forces are controlled as disturbance forces on hydraulic system by distributing it to each actuator through Jacobi matrix transformation. Decoupling control is given based on the dynamic model as well as a feed forward disturbance force compensation control strategy. However, due to the fact that differentiating acceleration which contains large noise is needed in decoupling control based on dynamic model, modal decoupling control is given. Modal equation of redundant shaking table is given by considering hydraulic cylinder as a hydraulic spring. Through standard modal matrix and its inverse matrix, the redundant shaking table is controlled in non-coupling modal space instead of DoFs space. By analyzing the relationship between the modal matrix and the coupling characteristics of different modal DoFs, an experimental method is given for determining the modal matrix. Simulation analysis shows that compared with decoupling control based on dynamic model, the modal space decoupling control can more effectively reduce the dynamic coupling among DoFs of the redundant shaking table. A control system of the redundant shaking table is developed using rapid control prototyping technology based on xPC Target. Detailed experimental analysis and research are carried out on the proposed coupling characteristic analysis and decoupling control strategies. Experimental results demonstrate that the proposed decoupling control strategies are effective and advanced.

J.-W. Han (✉) · Z.-D. Yang
Harbin Institute of Technology, Harbin, China
e-mail: hjw@fyty2010.com

Z.-D. Yang
e-mail: yangzhidong@hit.edu.cn

W. Wei
Institute of Machinery Manufacturing Technology,
China Academy of Engineering Physics, Mianyang, China
e-mail: 418159170@qq.com

V. Arakelian (ed.), *Dynamic Decoupling of Robot Manipulators*, Mechanisms and Machine Science 56, https://doi.org/10.1007/978-3-319-74363-9_5

5.1 Introduction

The existence of coupling not only results in the coupling error output of the parallel manipulator system, but also affects accuracy of the system. The coupling between the channels in the free space or the joint space makes it difficult to design the physical space channel controller of the multi input multi output parallel manipulator independently [1], which limits the integral control performance of the mechanism system. Paper [2–6] recognize the existence of coupling characteristics of parallel manipulators, but does not elaborate this coupling phenomenon and explain the essence of coupling explicitly. At present, there are many methods of evaluating the coupling extent of multivariable systems, and the static and dynamic relative gain matrix theory and the block relative gain matrix theory are the most widely used [7, 8]. These evaluation methods describe the relation between the input and output variables of the system, which ignore the internal information of the system and prone to coupling analysis bias [9]. Zhang et al. [10] analyzed the influence of inertia term of the dynamic model on the driving force.

Reasons for the dynamic coupling of parallel manipulator s mainly include the following aspects: the centroid position deviation of the moving platform and payload, the position change of the mechanism in the process of motion and the effect of branches. As for redundant multi-axial vibration table, due to its small motion range, the position and pose of the platform hardly change in the process of motion. Therefore, one can think the reason for the dynamic coupling of redundant multi-axial vibration table is centroid position deviation of the payload which results in the difference between comprehensive centroid point and control point. When the platform produces a certain degree of freedom motion, its inertia force (inertia moment) will produce the inertial moment (inertial force) relative to the center of mass, which makes the vibration table produce a coupling output of another degrees of freedom, that is, the motion of a certain degree of freedom affects other degrees of freedom. Three groups of hydraulic actuators are arranged perpendicular to each other, and zero linear degree control can be thought as an approximate decoupling control strategy when the platform is free of load due to the particular structure.

If the single cylinder control method is used as Stewart platform, the control system is also in the coupling state when the platform is free of load.

The reason for the dynamic coupling of redundant multi-axial vibration table is centroid position deviation of the payload, which makes the generalized mass of the system an off-diagonal matrix. Therefore, the basic idea of the dynamic decoupling control method is that the generalized mass matrix of the compensated dynamic equation is approximately diagonal through the appropriate compensation.

Based on this idea, there are several dynamic decoupling methods:

(1) Transient force decoupling. Centroid position deviation of the payload indicates that the dynamic coupling is generated by the generalized inertial force of

platform and the load, thus one of the methods available is that the generalized inertia force of the moving platform and the payload is distributed to each actuator to produce an additional disturbance force. Then, the actuator compensates the additional disturbance force, that is, the generalized inertial force is offset, and this method is called transient force decoupling. Paper [11] pointed out that the transient force is the inertial force of the payload which is reverse direction to the driving force due to the platform motion, the transient force will cause the rotational torque in the three rotational degrees of freedom. By compensating the transient force, the rotation direction coupling motion caused by translational motion is eliminated. Zhidong [12] established the linear model of three directional hydraulic vibration tables, then, proposes the feedback compensation strategy of acceleration proportional differential, whose validation was verified by simulation. As for six degrees of freedom MAST, Dr. Gizatullin [13] of University of Bath also calculated the overturning moment of the direction of rotation by the load acceleration, and controlled the overturning torque as disturbance force reducing the coupling between degrees of freedom.

(2) Space transformation decoupling. Based on the basic idea of dynamic decoupling, the decoupling control can be achieved by space transformation. There are two modes of transformation: mode transformation and singular value decomposition of generalized mass matrix. The common point of the two implementations is that the corresponding transformation matrix is generated to generate a new "space", so that the generalized mass matrix is diagonal matrix in the space. The different point is that the way of generating the transformation matrix. The generalized mass matrix diagonalization is realized by modal transformation based on the stiffness matrix. Singular value decomposition (SVD) is aimed at generalized mass matrices to realize the diagonalization of generalized mass matrix. Thus, the new "space" is also different, that is, the transformation matrix is different. It should be noted that gravity and Coriolis force and centrifugal force should be offset through the nonlinear compensation before space transformation decoupling is performed. Based on the principle that the vibration modes are orthogonal to the inertia matrix and the stiffness matrix [14], Plummer [15, 16] proposed the idea of modal decoupling control in the flight simulator control in 2007. The control of flight simulator usually adopts a single cylinder control, which transforms the control of the displacement of the DOF to that of the displacement of the single cylinder. But due to the strong coupling characteristic of the system in the joint space, the coupling exists among each cylinder controller, as a result, adjusting the parameters of a controller will affect all DOF dynamic characteristics. Plummer transforms the single cylinder control of the joint space into the uncoupled space by the modal matrix to realize the decoupling control of the flight simulator. Peter [17] and Yang [18, 19] apply the modal decoupling control to Stewart parallel mechanism with six degrees of freedom, which transforms multiple input multiple output (MIMO) system into multiple single input single

output (SISO) control system, and the validity of modal decoupling control was verified through the simulation and experiments. Jiang [20] introduced dynamic pressure feedback control strategy into modal space to improve the damping ratio of modal freedom. The bandwidth of the attitude freedom of the system is extended by increasing the dynamic characteristics of each mode, and the displacement of freedom degree disturbance caused by gravity is eliminated through the feedback compensation of gravity. The control strategy is shown in Fig. 5.1 [21], through singular value decomposition, transform the dynamic model of Stewart platform in joint space into the model in uncoupled space, and realize the independent design of each channel controller of Stewart platform.

In this work, transient force decoupling and modal space control will be applied to the dynamic coupling of the mechanism. Firstly, the space force system of redundant vibration table under eccentric load is analyzed, establishing the dynamic equations of the system, and the coupling mass matrix is given. Based on the coupled mass matrix, the coupling force observer is introduced. Through the Jacobi matrix transformation, the coupling force is decomposed to hydraulic actuators, here the coupling force is regarded as the disturbance force of the hydraulic system. Then, based on the analysis of disturbance force compensation of hydraulic system, the proportional differential compensation control strategy of disturbance force is proposed, therefore the decoupling control based on dynamic model is realized. But decoupling control based on dynamic model requires differential processing of acceleration feedback signal, therefore, based on the analysis of the dynamics equations, the decoupling control strategy is proposed due to the great noise of acceleration feedback signal in actual control. Vibration equation of the redundant vibration table will be analyzed, when the hydraulic cylinder is regarded as a hydraulic spring. Through the modal matrix and its inverse matrix, the freedom space of the redundant vibration table is transformed into its uncoupled modal space, then the modal decoupling control of the redundant vibration table is realized. Finally, the modal matrix of the eccentric load is analyzed, and the method of determining the modal matrix of redundant vibration table is given.

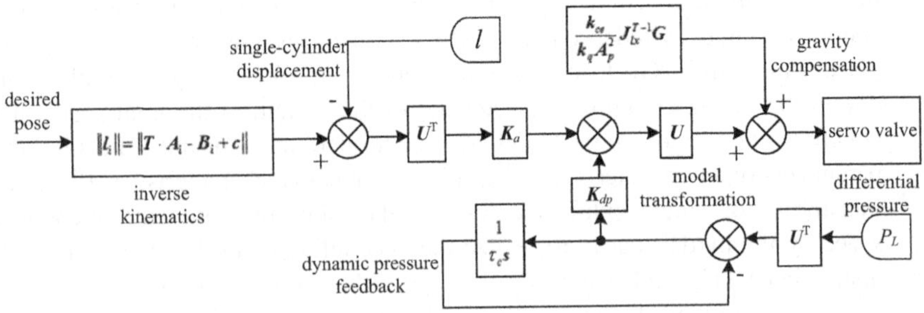

Fig. 5.1 Modal space control with gravity compensation

5.2 Transient Force Decoupling

5.2.1 Dynamic Coupling Analysis

Due to high bearing capacity of the redundant vibrating, and the mass of the cylinder and piston rod of hydraulic cylinder can be negligible compared with platform and load quality, thus, the dynamic model can be described by single rigid body dynamics. Based on the Newton-Euler, the dynamic equation can be given by [22, 23]

$$\boldsymbol{f}_q = \boldsymbol{M}(\boldsymbol{q})\dot{\boldsymbol{q}} + \boldsymbol{C}(\boldsymbol{q}, \dot{\boldsymbol{q}})\dot{\boldsymbol{q}} + \boldsymbol{G}(\boldsymbol{q}) \tag{5.1}$$

where,

$\boldsymbol{M}(\boldsymbol{q})$ Inertial mass matrix;
$\boldsymbol{C}(\boldsymbol{q}, \dot{\boldsymbol{q}})$ Centrifugal force and Coriolis force coefficient matrix;
$\boldsymbol{G}(\boldsymbol{q})$ Gravity vector;
$\boldsymbol{f}_q$ Driving force vector in freedom space.

Driving force vector can be given by

$$\boldsymbol{f}_q = [f_x \quad f_y \quad f_z \quad M_x \quad M_y \quad M_z]^{\mathrm{T}} \tag{5.2}$$

Due to the small amplitude high frequency motion of the vibration table, the effect of gravity and centrifugal force and the Coriolis on the dynamic characteristics of the system can be negligible. Thus, only the inertia force is considered to analyze the dynamic characteristic of redundant vibrating table. When the redundant vibration table is under load, the inertial force system of the system mainly includes the upper platform inertial force system $\{\boldsymbol{f}_{mo}\}$, the load inertia force system $\{\boldsymbol{f}_{ma}\}$ and the driving force system $\{\boldsymbol{f}_q\}$. In Fig. 5.2 Eccentric vector of load centroid is denoted by $\Delta\boldsymbol{b} = [\Delta x \quad \Delta y \quad \Delta z]^{\mathrm{T}}$. The action point of $\{\boldsymbol{f}_{mo}\}$ and $\{\boldsymbol{\tau}\}$ are at the centroid of platform o_p. The action point of $\{\boldsymbol{f}_{ma}\}$ is at the centroid of payload o_m.

Directions of inertial force system $\{\boldsymbol{f}_{ma}\}$ and $\{\boldsymbol{f}_{mo}\}$ are opposite to their respective acceleration directions, they can be given by

$$\boldsymbol{f}_{mo} = \left[m_o\ddot{x} \quad m_o\ddot{y} \quad m_o\ddot{z} \quad I_{mo,xp}\ddot{R}x \quad I_{mo,yp}\ddot{R}y \quad I_{mo,zp}\ddot{R}z\right]^{\mathrm{T}} \tag{5.3}$$

$$\boldsymbol{f}_{ma} = \left[m_a\ddot{x}_a \quad m_a\ddot{y}_a \quad m_a\ddot{z}_a \quad I_{ma,xm}\ddot{R}x \quad I_{ma,ym}\ddot{R}y \quad I_{ma,zm}\ddot{R}z\right]^{\mathrm{T}} \tag{5.4}$$

where,

m_o Mass of platform, (kg);
m_a Mass of load, (kg);
$I_{mo,ip}$ Moment of inertia of platform relative to x_p, y_p and z_p, (kg m^2);
$I_{ma,im}$ Moment of inertia of load relative to x_m, y_m and z_m, (kg m^2).

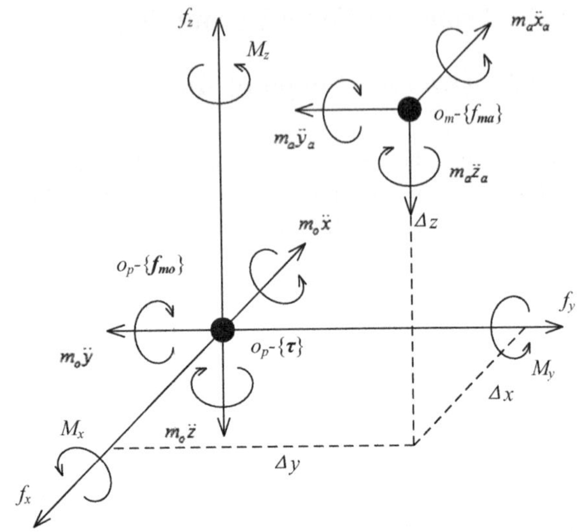

Fig. 5.2 Space force system with eccentric load

Based on the theory of rigid body dynamics, the relation between the load centroid acceleration and the platform centroid acceleration can be given by

$$\ddot{x}_{\mathrm{a}} = \ddot{x} + \Delta z\ddot{R}y - \Delta y\ddot{R}z \tag{5.5}$$

$$\ddot{y}_{\mathrm{a}} = \ddot{y} - \Delta z\ddot{R}x + \Delta x\ddot{R}z \tag{5.6}$$

$$\ddot{z}_a = \ddot{z} + \Delta y\ddot{R}x - \Delta x\ddot{R}y \tag{5.7}$$

According to the Fig. 5.2, force balance equations in different degrees of freedom can be obtained

$$f_x = m_o\ddot{x} + m_a\ddot{x}_a \tag{5.8}$$

$$f_y = m_o\ddot{y} + m_a\ddot{y}_a \tag{5.9}$$

$$f_z = m_o\ddot{z} + m_a\ddot{z}_a \tag{5.10}$$

$$M_x = I_{mo,xp}\ddot{R}x + I_{ma,xm}\ddot{R}x - m_a\ddot{y}_a\Delta z + m_a\ddot{z}_a\Delta y \tag{5.11}$$

$$M_y = I_{mo,yp}\ddot{R}y + I_{ma,ym}\ddot{R}y + m_a\ddot{x}_a\Delta z - m_a\ddot{z}_a\Delta x \tag{5.12}$$

$$M_z = I_{mo,zp}\ddot{R}z + I_{ma,zm}\ddot{R}z - m_a\ddot{x}_a\Delta y + m_a\ddot{y}_a\Delta x \tag{5.13}$$

Substituting Eqs. (5.5) and (5.6) into Eqs. (5.8)–(5.13), the following equation can be given

$$f_x = m_o\ddot{x} + m_a\ddot{x} + m_a\Delta z\ddot{R}y - m_a\Delta y\ddot{R}z \tag{5.14}$$

$$f_y = m_o\ddot{y} + m_a\ddot{y} - m_a\Delta z\ddot{R}x + m_a\Delta x\ddot{R}z \tag{5.15}$$

$$f_z = m_o\ddot{z} + m_a\ddot{z} + m_a\Delta y\ddot{R}x - m_a\Delta x\ddot{R}y \tag{5.16}$$

$$\begin{aligned} M_x = & -m_a\Delta z\ddot{y} + m_a\Delta y\ddot{z} + \\ & + \left[I_{mo,xp} + I_{ma,xm} + m_a\left(\Delta y^2 + \Delta z^2\right)\right]\ddot{R}x - m_a\Delta x\Delta y\ddot{R}y - m_a\Delta x\Delta z\ddot{R}z \end{aligned} \tag{5.17}$$

$$\begin{aligned} M_y = & \, m_a\Delta z\ddot{x} - m_a\Delta x\ddot{z} + \\ & - m_a\Delta x\Delta y\ddot{R}x + \left[I_{mo,yp} + I_{ma,ym} + m_a\left(\Delta x^2 + \Delta z^2\right)\right]\ddot{R}y - m_a\Delta y\Delta z\ddot{R}z \end{aligned} \tag{5.18}$$

$$\begin{aligned} M_z = & -m_a\Delta y\ddot{x} + m_a\Delta x\ddot{y} + \\ & - m_a\Delta x\Delta z\ddot{R}x - m_a\Delta y\Delta z\ddot{R}y + \left[I_{mo,zp} + I_{ma,zm} + m_a\left(\Delta x^2 + \Delta y^2\right)\right]\ddot{R}z \end{aligned} \tag{5.19}$$

Noting that moment of inertia of load relative to x_p, y_p and z_p can be given by

$$\begin{cases} I_{ma,xp} = I_{ma,xm} + m_a(\Delta y^2 + \Delta z^2) \\ I_{ma,yp} = I_{ma,ym} + m_a(\Delta x^2 + \Delta z^2) \\ I_{ma,zp} = I_{ma,zm} + m_a(\Delta x^2 + \Delta y^2) \end{cases} \tag{5.20}$$

Defining the total mass as $m = m_o + m_a$, and defining the total inertia as $I_i = I_{mo,ip} + I_{ma,ip}$, Eqs. (5.14)–(5.19) can be described by

$$\boldsymbol{f_q} = \hat{\boldsymbol{M}}\ddot{\boldsymbol{q}} + \boldsymbol{M_c}\ddot{\boldsymbol{q}} \tag{5.21}$$

where, diagonal inertial matrix $\hat{\boldsymbol{M}}$ can be given

$$\hat{\boldsymbol{M}} = \mathrm{diag}(m, m, m, I_x, I_y, I_z) \tag{5.22}$$

The coupled inertial matrix can be given by

$$\boldsymbol{M_c} = \begin{bmatrix} 0 & 0 & 0 & 0 & m_a\Delta z & -m_a\Delta y \\ 0 & 0 & 0 & -m_a\Delta z & 0 & m_a\Delta x \\ 0 & 0 & 0 & m_a\Delta y & -m_a\Delta x & 0 \\ 0 & -m_a\Delta z & m_a\Delta y & 0 & -m_a\Delta x\Delta y & -m_a\Delta x\Delta z \\ m_a\Delta z & 0 & -m_a\Delta x & -m_a\Delta x\Delta y & 0 & -m_a\Delta y\Delta z \\ -m_a\Delta y & m_a\Delta x & 0 & -m_a\Delta x\Delta z & -m_a\Delta y\Delta z & 0 \end{bmatrix}$$

In the Eq. (5.21), $\boldsymbol{M_c}\ddot{\boldsymbol{q}}$ denotes the dynamic coupling section. From the coupling matrix $\boldsymbol{Mc}$, it can be seen that there exists the coupling between the degrees of freedom of the redundant vibrating table. The load eccentricity of the redundant vibrating table is mainly in the vertical direction, $\Delta x = \Delta y = 0$. Here, the coupling of the system is represented by the coupling between the X translation and the Ry rotation and the coupling between the Y translation and the Rx rotation. Thus, the coupling matrix can be simplified

$$\boldsymbol{M}_c = \begin{bmatrix} 0 & 0 & 0 & 0 & m_a\Delta z & 0 \\ 0 & 0 & 0 & -m_a\Delta z & 0 & 0 \\ 0 & 0 & 0 & 0 & 0 & 0 \\ 0 & -m_a\Delta z & 0 & 0 & 0 & 0 \\ m_a\Delta z & 0 & 0 & 0 & 0 & 0 \\ 0 & 0 & 0 & 0 & 0 & 0 \end{bmatrix} \tag{5.23}$$

Due to the equal work in freedom space and in driving space of the hydraulic cylinder, one can obtain

$$\boldsymbol{f}_q^{\mathrm{T}}\dot{\boldsymbol{q}} = \boldsymbol{f}_l^{\mathrm{T}}\dot{\boldsymbol{l}} \tag{5.24}$$

Thus, one can obtain the relationship between force vector in freedom space $\boldsymbol{f}_q$ and that in joint space $\boldsymbol{f}_l$, which can be given by

$$\boldsymbol{f}_q = \boldsymbol{J}^{\mathrm{T}}(\boldsymbol{q})\boldsymbol{f}_l \tag{5.25}$$

Combining Eqs. (5.21) and (5.25), one can obtain the dynamic equation in zero-position

$$\boldsymbol{C}^{\mathrm{T}}\boldsymbol{f}_l = \left(\hat{\boldsymbol{M}} + \boldsymbol{M}_c\right)\ddot{\boldsymbol{q}} \tag{5.26}$$

5.2.2 *Decoupling Control Based on Dynamic Model*

5.2.2.1 Feed Forward Compensation Control for Disturbing Force

Furthermore, Eq. (5.26) can be written as

$$C^{\mathrm{T}}\left[f_l - (C^{\mathrm{T}})^{+}M_c\ddot{q}\right] = \hat{M}\ddot{q} \tag{5.27}$$

where, $(\boldsymbol{C}^{\mathrm{T}})^{+} = (\boldsymbol{C}\boldsymbol{C}^{\mathrm{T}})^{-1}\boldsymbol{C}$.

Defining $\boldsymbol{f}_d = (\boldsymbol{C}^{\mathrm{T}})^{+}\boldsymbol{M}_c\ddot{\boldsymbol{q}}$. Thus, $\boldsymbol{f}_d$ is the interference force of each hydraulic driving system to which freedom degree coupling force is decomposed. For a single electro-hydraulic servo system, the block diagram of the transfer function with the interference force is expressed in Fig. 5.3.

The transfer function between the displacement l and the given signal u and the interference force f_d can be given by

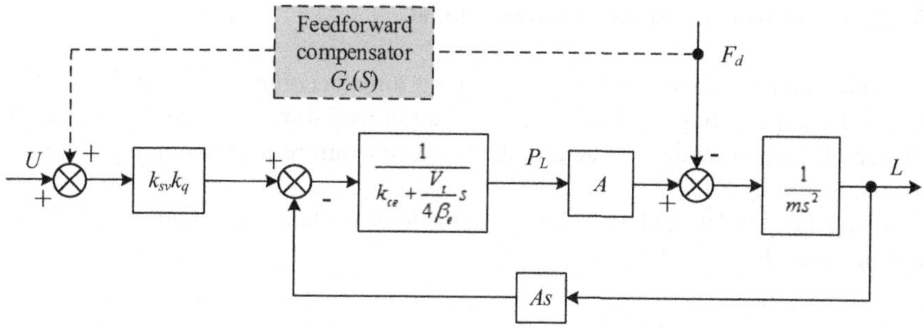

Fig. 5.3 Disturbance force control of hydraulic system

$$L = G_u(s)U - G_d(s)F_d \tag{5.28}$$

where,

$$G_u(s) = \frac{k_{sv}k_q}{As\left(\frac{s^2}{\omega_h^2} + \frac{2\varsigma_h}{\omega_h}s + 1\right)};$$

$$G_d(s) = \frac{\frac{k_{ce}}{A^2}\left(\frac{V_t}{4\beta_e k_{ce}}s + 1\right)}{s\left(\frac{s^2}{\omega_h^2} + \frac{2\varsigma_h}{\omega_h}s + 1\right)}.$$

After feed forward controller for the disturbance force $G_c(s)$ added, the transfer function between the displacement l and the given signal u and the disturbance force f_d becomes

$$\tilde{L} = G_u(s)[U + G_c(s)F_d] - G_d(s)F_d \tag{5.29}$$

In order to make the output displacement free from the influence of the disturbance force, feed forward controller for the disturbance force $G_d(s)$ should be obtained

$$G_c(s) = G_u(s)^{-1}G_d(s) = \frac{V_t}{4\beta_e k_{sv}k_qA}s + \frac{k_{ce}}{k_{sv}k_qA} \tag{5.30}$$

Equation (5.30) indicates that the feed forward compensation controller for the disturbance force is a proportional differential element. When the proportion and the differential coefficient satisfy Eq. (5.30), the interference force can be completely compensated.

5.2.2.2 Control Based on Dynamic Model

When the coupling mass matrix, the coupling force decomposition matrix and feed forward compensation controller for the interference force are introduced into the degree of freedom control structure, the block diagram of the decoupling control is shown in Fig. 5.4.

Using Eqs. (5.19) and (5.21), the output force of the single hydraulic cylinder can be given by

$$f_l = \frac{k_{sv}k_qA}{\frac{V_t}{4\beta_e}s + k_{ce}}u - \frac{A^2s}{\frac{V_t}{4\beta_e}s + k_{ce}}l = G_{fu}(s)u - G_{fl}(s)l \tag{5.31}$$

Assuming that each hydraulic driving system has the same dynamic characteristics, combing Eqs. (5.26) and (5.31), one can obtain

$$\hat{\boldsymbol{M}}\ddot{\boldsymbol{q}} + \boldsymbol{M}_c\ddot{\boldsymbol{q}} = C^T[G_{fu}(s)\boldsymbol{u} - G_{fl}(s)\boldsymbol{l}] \tag{5.32}$$

From Fig. 5.4, the output vector of decoupling control $\boldsymbol{u_d}$ can be given by

$$\boldsymbol{u_d} = G_d(s)(\boldsymbol{C}^{\mathrm{T}})^+\boldsymbol{M}_c\ddot{\boldsymbol{q}} \tag{5.33}$$

Substituting equations $\boldsymbol{u} = \boldsymbol{c}\boldsymbol{u_q} + \boldsymbol{u}_d$ and $\boldsymbol{l} = \boldsymbol{C}\boldsymbol{q}$ into Eq. (5.32), one can obtain

$$\hat{\boldsymbol{M}}\ddot{\boldsymbol{q}} + \boldsymbol{M}_c\ddot{\boldsymbol{q}} = G_{fu}(s)\boldsymbol{C}^{\mathrm{T}}\boldsymbol{C}\boldsymbol{u_q} + G_{fu}(s)G_d(s)(\boldsymbol{C}^{\mathrm{T}})^+\boldsymbol{M}_c\ddot{\boldsymbol{q}} - G_{fl}(s)\boldsymbol{C}^{\mathrm{T}}\boldsymbol{C}\boldsymbol{q} \tag{5.34}$$

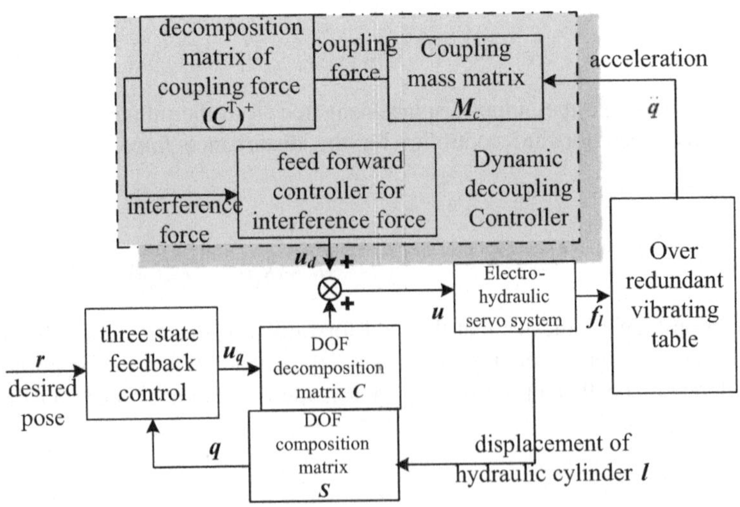

Fig. 5.4 Dynamic decoupling control diagram of redundant shaking table

Noting $G_{fu}(s)G_d(s) = 1$, $\boldsymbol{C}^{\mathrm{T}}(\boldsymbol{C}^{\mathrm{T}})^{+} = \boldsymbol{I}_{6\times6}$, equation **Erreur! Source du renvoi introuvable.** can also be written as

$$\hat{\boldsymbol{M}}\ddot{\boldsymbol{q}} + G_{fl}(s)\boldsymbol{C}^{\mathrm{T}}\boldsymbol{C}\boldsymbol{q} = G_{fu}(s)\boldsymbol{C}^{T}\boldsymbol{C}\boldsymbol{u}_{\boldsymbol{q}} \tag{5.35}$$

Considering the structural characteristics of the redundant vibrating table, the matrix $\boldsymbol{C}^{\mathrm{T}}\boldsymbol{C}$ is a diagonal matrix in zero position

$$\boldsymbol{C}^{\mathrm{T}}\boldsymbol{C} = \begin{bmatrix} 4 & 0 & 0 & 0 & 0 & 0 \\ 0 & 4 & 0 & 0 & 0 & 0 \\ 0 & 0 & 4 & 0 & 0 & 0 \\ 0 & 0 & 0 & 4L_z^2 & 0 & 0 \\ 0 & 0 & 0 & 0 & 4L_z^2 & 0 \\ 0 & 0 & 0 & 0 & 0 & 2L_{x1}^2 + 2L_{x2}^2 + 2L_{y1}^2 + 2L_{x2}^2 \end{bmatrix} \tag{5.36}$$

Thus, in Eq. (5.35), there is no coupling between the degree of freedom $\boldsymbol{q}$ and the control quantity $\boldsymbol{u}_{\boldsymbol{q}}$.

For the redundant vibration table, the multiple input multiple output (MIMO) system is transformed into multiple single input single output (SISO) systems through decoupling controller, and independent control of each degree of freedom can be achieved.

5.3 Modal Decoupling

5.3.1 *Free Vibration Model of Redundant Vibrating Table*

The hydraulic cylinder in the redundant hydraulic vibration table is considered as a hydraulic spring, and the free vibration model can be obtained just as shown in Fig. 5.5, where the stiffness of the hydraulic spring is

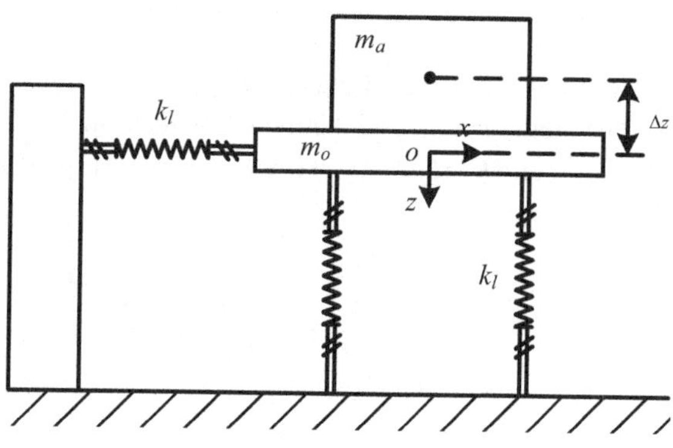

Fig. 5.5 Free vibration model of redundant shaking table

$$k_l = \frac{4\beta_e A^2}{V_t} \tag{5.37}$$

The vibration differential equation of the free vibration model can be expressed as:

$$M\ddot{q} + Kq = 0 \tag{5.38}$$

Combing the Eqs. (5.22) and (5.23), the mass matrix $\boldsymbol{M}$ is

$$\boldsymbol{M} = \begin{bmatrix} m & 0 & 0 & 0 & m_a\Delta z & 0 \\ 0 & m & 0 & -m_a\Delta z & 0 & 0 \\ 0 & 0 & m & 0 & 0 & 0 \\ 0 & -m_a\Delta z & 0 & I_x & 0 & 0 \\ m_a\Delta z & 0 & 0 & 0 & I_y & 0 \\ 0 & 0 & 0 & 0 & 0 & I_z \end{bmatrix} \tag{5.39}$$

The stiffness matrix $\boldsymbol{K}$ is expressed as

$$\boldsymbol{K} = k_l \boldsymbol{C}^{\mathrm{T}}\boldsymbol{C} \tag{5.40}$$

The vibration modes are orthogonal to the mass matrix and the stiffness matrix according to the vibration modal theory [14], so the modal matrix $\boldsymbol{\Phi}$ exists, that

$$\tilde{\boldsymbol{M}} = \boldsymbol{\Phi}^{\mathrm{T}}\boldsymbol{M}\boldsymbol{\Phi} \tag{5.41}$$

$$\tilde{\boldsymbol{K}} = \boldsymbol{\Phi}^{\mathrm{T}}\boldsymbol{K}\boldsymbol{\Phi} \tag{5.42}$$

where $\tilde{\boldsymbol{M}}$ is the modal mass matrix, $\tilde{\boldsymbol{K}}$ is the modal stiffness matrix.

$$\tilde{\boldsymbol{K}} = \begin{bmatrix} \omega_1^2 & 0 & \cdots & 0 \\ 0 & \omega_2^2 & \cdots & 0 \\ \vdots & \vdots & \vdots & \vdots \\ 0 & 0 & \cdots & \omega_6^2 \end{bmatrix} \tilde{\boldsymbol{M}} \tag{5.43}$$

where ω_j is the natural frequency corresponding to the rank j.

The modal matrix and the modal stiffness matrix can be solved by the statement $[\boldsymbol{\Phi}, \tilde{\boldsymbol{K}}] = \mathrm{eig}(\boldsymbol{K}, \boldsymbol{M})$ in MATLAB. Combing Eqs. (5.40) and (5.42), we can obtain that

$$\boldsymbol{\Phi}^{\mathrm{T}}\boldsymbol{C}^{\mathrm{T}}\boldsymbol{C}\boldsymbol{\Phi} = \frac{\tilde{\boldsymbol{K}}}{k_l} \tag{5.44}$$

5.3.2 Modal Space Decoupling Control

Defining $\boldsymbol{c}_r = \frac{V_t}{4\beta_e}$, and taking the vibration table of the 12 hydraulic cylinders into account, then the Eq. (5.31) can be rewritten as

$$f_l = \frac{k_{sv}k_qA}{c_rs + k_{ce}}\boldsymbol{u} - \frac{A^2s}{c_rs + k_{ce}}\boldsymbol{l} \tag{5.45}$$

Substituting Eq. (5.45) to Eq. (5.46), and performing the Laplace transform, then we can get

$$\boldsymbol{M}s^2\boldsymbol{q} = \boldsymbol{C}^{\mathrm{T}}\left(\frac{k_{sv}k_qA}{c_rs + k_{ce}}\boldsymbol{u} - \frac{A^2s}{c_rs + k_{ce}}\boldsymbol{l}\right) \tag{5.46}$$

Defining $\boldsymbol{u} = \boldsymbol{C}\boldsymbol{u}_q$, then Eq. (5.46) can be rewritten as

$$\boldsymbol{M}s^2\boldsymbol{q} = \frac{k_{sv}k_qA}{c_rs + k_{ce}}\boldsymbol{C}^{\mathrm{T}}\boldsymbol{C}\boldsymbol{u}_q - \frac{A^2s}{c_rs + k_{ce}}\boldsymbol{C}^{\mathrm{T}}\boldsymbol{C}\boldsymbol{q} \tag{5.47}$$

Combing Eqs. (5.37) and (5.46), we can obtain that

$$s\left(\frac{\boldsymbol{M}}{k_l}s^2 + \frac{\boldsymbol{M}k_{ce}}{A^2}s + \boldsymbol{C}^{\mathrm{T}}\boldsymbol{C}\right)\boldsymbol{q} = k_{sh}\boldsymbol{C}^{\mathrm{T}}\boldsymbol{C}\boldsymbol{u}_q \tag{5.48}$$

Therefore the Eq. (5.48) is the system transfer function with zero degree of freedom linearization control. There exists a coupling between the degrees of freedom of the system and the position and orientation of the system since the mass matrix $\boldsymbol{M}$ is a non-diagonal matrix. Then Introducing the standard modal matrix $\boldsymbol{\Phi}$, and defining $\boldsymbol{q} = \boldsymbol{\Phi}\tilde{\boldsymbol{q}}$, $\boldsymbol{u}_{\mathrm{q}} = \boldsymbol{\Phi}\tilde{\boldsymbol{u}}$, Eq. (5.48) can be rewritten as

$$s\boldsymbol{\Phi}^{\mathrm{T}}\left(\frac{\boldsymbol{M}}{k_l}s^2 + \frac{\boldsymbol{M}k_{ce}}{A^2}s + \boldsymbol{C}^{\mathrm{T}}\boldsymbol{C}\right)\boldsymbol{\Phi}\tilde{\boldsymbol{q}} = k_{sh}\boldsymbol{\Phi}^{\mathrm{T}}\boldsymbol{C}^{\mathrm{T}}\boldsymbol{C}\boldsymbol{\Phi}\tilde{\boldsymbol{u}} \tag{5.49}$$

Simplify as

$$s\left(\frac{\tilde{\boldsymbol{M}}}{k_l}s^2 + \frac{\tilde{\boldsymbol{M}}k_{ce}}{A^2}s + \frac{\tilde{\boldsymbol{K}}}{k_l}\right)\tilde{\boldsymbol{q}} = \frac{k_{sh}}{k_l}\tilde{\boldsymbol{K}}\tilde{\boldsymbol{u}} \tag{5.50}$$

Since the modal mass matrix $\tilde{\boldsymbol{M}}$ and the modal stiffness matrix $\tilde{\boldsymbol{K}}$ are all diagonal matrices, the system can be decoupled from MIMO into 6 SISO three order systems after the modal matrix transformation, and the independent control can be achieved between the modal displacement and the modal control. The open-loop transfer function of order $\boldsymbol{i}$ can be described as

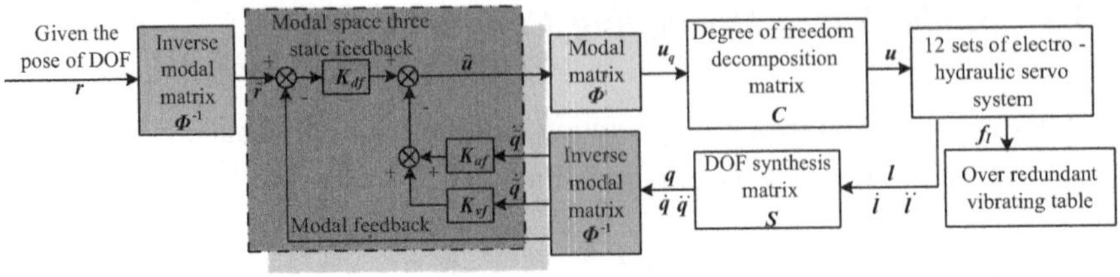

Fig. 5.6 Modal decoupling control of redundant shaking table

$$s\left(\frac{s^2}{\omega_i^2}+\frac{s}{A^2\omega_i^2/k_{ce}k_l}+1\right)\tilde{q}_i=k_{sh}\tilde{u}_i \tag{5.51}$$

where ω_i is the natural frequency of order $\boldsymbol{i}$ modes, $\tilde{\boldsymbol{q}}_i$ is the modal displacement of order $\boldsymbol{i}$, $\hat{\boldsymbol{u}}_i$ is the modal control quantity of order $\boldsymbol{i}$. According to the Eq. (5.51), we can obtain that

$$2\frac{\omega_i}{\zeta_i}=\frac{A^2\omega_i^2}{k_{ce}k_l} \tag{5.52}$$

where ζ_i is the damping ratio of the order $\boldsymbol{i}$ modal. By Eq. (5.52), we can obtain that

$$\zeta_i=\frac{k_{ce}k_l}{2A^2\omega_i} \tag{5.53}$$

According to the above analysis, the three state control strategy is introduced into the modal space for the control, and we can obtain the modal decoupling control strategy as shown in Fig. 5.6. The displacement, velocity and acceleration feedback signals of hydraulic cylinders are transformed into degrees of freedom feedback signals through the degrees of freedom synthesis matrix, and the feedback signals of degrees' freedom are transformed into modal feedback signals by modal changes, and three state feedback control is involved in modal space. The modal control quantity $\tilde{\boldsymbol{u}}$ is transformed into the input voltage $\boldsymbol{u}$ of the 12 way electro-hydraulic servo valve through the modal matrix and the degree of freedom decomposition matrix, so that the decoupling control of redundant vibration table in modal space can be realized.

5.3.3 *Modal Matrix Analysis of Load Eccentricity*

Although the above gives the calculation method of modal matrix, but in the actual system, load inertia, eccentricity and the equivalent spring stiffness of the hydraulic cylinder is very difficult to accurately determine, so it is necessary to analyze the modal matrix and give its simple experimental determination method.

When the load is only vertical eccentric, the coupling of the redundant degrees of freedom of the vibration table is mainly the coupling between $\boldsymbol{x}$ and $\boldsymbol{Ry}$, and between $\boldsymbol{y}$ and $\boldsymbol{Rx}$. The coupling matrix is analyzed by the coupling between $\boldsymbol{x}$ and $\boldsymbol{Ry}$, the coupling mass matrix $\boldsymbol{M}_{\boldsymbol{x},\boldsymbol{Ry}}$ and matrix $\left(\boldsymbol{C}^{\mathrm{T}}\boldsymbol{C}\right)_{x,Ry}$ between $\boldsymbol{x}$ and $\boldsymbol{Ry}$ can be obtained by the Eqs. (5.36) and (5.39), respectively as

$$\boldsymbol{M}_{\boldsymbol{x},\boldsymbol{Ry}} = \begin{bmatrix} m & m_a\Delta z \\ m_a\Delta z & I_y \end{bmatrix} \tag{5.54}$$

$$\left(\boldsymbol{C}^{\mathrm{T}}\boldsymbol{C}\right)_{x,Ry} = \begin{bmatrix} 4 & 0 \\ 0 & 4L_z^2 \end{bmatrix} \tag{5.55}$$

Substituting Eqs. (5.54) and (5.55) into Eq. (5.48), and we can obtain that

$$\begin{aligned} &\left(\frac{s^2}{k_l}\begin{bmatrix} m & m_a\Delta z \\ m_a\Delta z & I_y \end{bmatrix} + \frac{k_{ce}s}{A^2}\begin{bmatrix} m & m_a\Delta z \\ m_a\Delta z & I_y \end{bmatrix} + \begin{bmatrix} 4 & 0 \\ 0 & 4L_z^2 \end{bmatrix}\right)\begin{bmatrix} \dot{x} \\ \dot{R}y \end{bmatrix} \\ &= k_{sh}\begin{bmatrix} 4 & 0 \\ 0 & 4L_z^2 \end{bmatrix}\begin{bmatrix} u_x \\ u_{Ry} \end{bmatrix} \end{aligned} \tag{5.56}$$

And the Eq. (5.56) can be rewritten as

$$\begin{cases} \left(\frac{ms^2}{k_l} + \frac{mk_{ce}s}{A^2} + 4\right)\dot{x} + \left(\frac{m_a\Delta zs^2}{k_l} + \frac{m_a\Delta zk_{ce}s}{A^2}\right)\dot{R}y = 4k_{sh}u_x \\ \left(\frac{m_a\Delta zs^2}{k_l} + \frac{m_a\Delta zk_{ce}s}{A^2}\right)\dot{x} + \left(\frac{I_ys^2}{k_l} + \frac{I_yk_{ce}s}{A^2} + 4L_z^2\right)\dot{R}y = 4L_z^2k_{sh}u_{Ry} \end{cases} \tag{5.57}$$

Defining:

$$\begin{aligned} G_x &= \frac{ms^2}{k_l} + \frac{mk_{ce}s}{A^2} + 4; \\ G_{Ry} &= \frac{I_ys^2}{k_l} + \frac{I_yk_{ce}s}{A^2} + 4L_z^2; \\ G_{\Delta z} &= \frac{m_a\Delta zs^2}{k_l} + \frac{m_a\Delta zk_{ce}s}{A^2}. \end{aligned}$$

So the Eq. (5.57) can be expressed as a matrix form

$$\begin{bmatrix} G_x & G_{\Delta z} \\ G_{\Delta z} & G_{Ry} \end{bmatrix}\begin{bmatrix} \dot{x} \\ \dot{R}y \end{bmatrix} = k_{sh}\begin{bmatrix} 4 & 0 \\ 0 & 4L_z^2 \end{bmatrix}\begin{bmatrix} u_x \\ u_{Ry} \end{bmatrix} \tag{5.58}$$

At the end of Eq. (5.58) equals sign, multiply left by

$$\begin{bmatrix} G_x & G_{\Delta z} \\ G_{\Delta z} & G_{Ry} \end{bmatrix}^{-1} = \frac{1}{G_xG_{Ry} - G_{\Delta z}^2}\begin{bmatrix} G_{Ry} & -G_{\Delta z} \\ -G_{\Delta z} & G_x \end{bmatrix} \tag{5.59}$$

And we can obtain that

$$\begin{bmatrix} \dot{x} \\ \dot{R}y \end{bmatrix} = \frac{4k_{sh}}{G_x G_{Ry} - G_{\Delta z}^2} \begin{bmatrix} G_{Ry} & -L_z^2 G_{\Delta z} \\ -G_{\Delta z} & L_z^2 G_x \end{bmatrix} \begin{bmatrix} u_x \\ u_{Ry} \end{bmatrix} \tag{5.60}$$

From the Eq. (5.60), we can see the coupling relation between $\boldsymbol{x}$ and $\boldsymbol{Ry}$ under the degree of freedom control strategy. The non-diagonal element $-\boldsymbol{L}_z^2 \boldsymbol{G}_{\Delta z}$ in the matrix represents the degree of coupling of the $\boldsymbol{x}$ translation direction when the $\boldsymbol{Ry}$ rotates; the non-diagonal element $-\boldsymbol{G}_{\Delta z}$ reflects the degree of coupling of the rotation direction of the $\boldsymbol{Ry}$ in $\boldsymbol{x}$ translation. $\boldsymbol{L}_z$ is related to the structural arrangement of the redundant vibrating table. The redundant vibration table structure in this work is $\boldsymbol{L}_z = 0.5\,\text{m}$, and $-\boldsymbol{G}_{\Delta z}$ is 4 times of $-\boldsymbol{L}_z^2 \boldsymbol{G}_{\Delta z}$ under this structure. This shows that the $\boldsymbol{Ry}$ rotation leads to smaller $\boldsymbol{x}$ direction translation coupling, while the $\boldsymbol{Ry}$ coupling caused by $\boldsymbol{x}$ translation is more rotational.

The modal matrix $\boldsymbol{\Phi}_{x,Ry}$ between $\boldsymbol{x}$ and $\boldsymbol{Ry}$ is represented as:

$$\boldsymbol{\Phi}_{x,Ry} = \begin{bmatrix} 1 & \varphi_2 \\ \varphi_1 & 1 \end{bmatrix} \tag{5.61}$$

The modal inverse matrix $\boldsymbol{\Phi}_{x,Ry}^{-1}$ is

$$\boldsymbol{\Phi}_{x,Ry}^{-1} = \frac{1}{1 - \varphi_1 \varphi_2} \begin{bmatrix} 1 & -\varphi_2 \\ -\varphi_1 & 1 \end{bmatrix} \tag{5.62}$$

As shown in Fig. 5.6, the relation between modal displacement $\tilde{\boldsymbol{q}}_{x,Ry}$ and modal control quantity $\tilde{\boldsymbol{u}}_{x,Ry}$ is obtained under the modal control strategy:

$$\begin{bmatrix} \dot{\tilde{x}} \\ \dot{\tilde{R}}y \end{bmatrix} = \frac{4k_{sh}}{G_x G_{Ry} - G_{\Delta z}^2} \boldsymbol{\Phi}_{x,Ry}^{-1} \begin{bmatrix} G_{Ry} & -L_z^2 G_{\Delta z} \\ -G_{\Delta z} & L_z^2 G_x \end{bmatrix} \boldsymbol{\Phi}_{x,Ry} \begin{bmatrix} \tilde{u}_x \\ \tilde{u}_{Ry} \end{bmatrix} \tag{5.63}$$

Substituting the Eq. (5.63) into Eqs. (5.61) and (5.62), and we can obtain

$$\begin{aligned} \begin{bmatrix} \dot{\tilde{x}} \\ \dot{\tilde{R}}y \end{bmatrix} = & \begin{bmatrix} G_{Ry} - L_z \varphi_1 \varphi_2 G_x + (\varphi_2 - L_z \varphi_1) G_{\Delta z} & \varphi_2 G_{Ry} - \varphi_2 L_z G_x + (\varphi_2^2 - L_z) G_{\Delta z} \\ \varphi_1 L_z G_x - \varphi_1 G_{Ry} + (\varphi_1^2 - 1) G_{\Delta z} & L_z G_x - \varphi_1 \varphi_2 G_{Ry} + (L_z \varphi_1 - \varphi_2) G_{\Delta z} \end{bmatrix} \\ & \times \frac{4k_{sh}}{(1 - \varphi_1 \varphi_2)(G_x G_{Ry} - G_{\Delta z}^2)} \begin{bmatrix} \tilde{u}_x \\ \tilde{u}_{Ry} \end{bmatrix} \end{aligned} \tag{5.64}$$

The non-diagonal element $\boldsymbol{\varphi}_2 \boldsymbol{G}_{\boldsymbol{Ry}} - \boldsymbol{\varphi}_2 \boldsymbol{L}_z \boldsymbol{G}_x + (\boldsymbol{\varphi}_2^2 - \boldsymbol{L}_z) \boldsymbol{G}_{\Delta z}$ in Eq. (5.64) is only associated with $\boldsymbol{\varphi}_2$, the term indicates the degree of coupling motion in the $\tilde{\boldsymbol{x}}$ direction when moving in the $\tilde{\boldsymbol{R}}\boldsymbol{y}$ direction, when the $\boldsymbol{\varphi}_2$ corresponds to the value in the modal matrix, the element will be 0, and the motion in the $\tilde{\boldsymbol{R}}\boldsymbol{y}$ direction will not cause the $\tilde{\boldsymbol{x}}$ direction to have coupled outputs. However, the non-diagonal element $\boldsymbol{\varphi}_1 \boldsymbol{L}_z \boldsymbol{G}_x - \boldsymbol{\varphi}_1 \boldsymbol{G}_{\boldsymbol{Ry}} + (\boldsymbol{\varphi}_1^2 - \boldsymbol{1}) \boldsymbol{G}_{\Delta z}$ is only associated with $\boldsymbol{\varphi}_1$, and the non-diagonal element is the degree of coupling motion of the $\tilde{Ry}$ in the $\tilde{x}$ motion, when the φ_1 corresponds to the value in the modal matrix, the value will be 0, and the motion of the $\tilde{x}$ will not cause the $\tilde{Ry}$ to have coupled outputs.

According to the above characteristics of φ_1 and φ_2, the modal matrix can be determined by the following experimental debugging methods: $\tilde{x}$ is given a sinusoidal signal, the $\tilde{Ry}$ has a corresponding sinusoidal output coupling, adjusting the value of φ_1, observe the coupled sine amplitude changes of $\tilde{Ry}$, if the coupling amplitude change is in the opposite direction to adjust when the φ_1 value is adjusted to a value corresponding to the modal matrix, the minimum amplitude of the output coupling $\tilde{Ry}$; Similarly, $\tilde{Ry}$ is given a sinusoidal signal, the $\tilde{x}$ has a corresponding sinusoidal output coupling, adjusting the value of φ_2, observe the coupled sine amplitude changes of $\tilde{x}$, if the change is in the opposite direction to adjust when the φ_2 value is adjusted to a value corresponding to the modal matrix, $\tilde{x}$ minimum output coupling.

5.4 Experiment

The hardware system of the redundant multi-axis vibrating table is shown in Fig. 5.7. It is composed of the servo value, hydraulic cylinder, upper platform, base, reaction wall, displacement sensor, acceleration sensor and pressure sensor etc.

The control system of the redundant multi-axis vibrating table is achieved using up-and low-level computer. The upper level computer is the general PC computer, including human-computer interaction interface (HCII) and data processing program established by LabVIEW. The low level computer is YanHua industry control used for running Real time servo control program. The RTSP is developed by rapid

Fig. 5.7 Redundant multi-axis shaking table

Table 5.1 Load experiment parameters of redundant shaking table

Parameters	Value	Parameters	Value
m_o	1500 kg	m_a	1500 kg
Δz	−0.7 m	I_x	950 kg × m^2
I_y	950 kg × m^2	I_z	550 kg × m^2

prototyping control technology based on xPC Target. In the upper level computer, the control prototype of the system is built with Simulink, and the control prototype is translated into real-time operation code by the compiler, which is loaded into the lower computer through the TCP/IP network protocol. The control cycle of the low level computer is 1 ms.

Since both LabVIEW and Simulink are graphical software development, the program runs in a clear procedure thus, they can significantly shorten software development time. The payload test parameters in the mechanical dynamics model are shown in Table 5.1.

5.4.1 Experiment of Dynamics Coupling Analysis

When the redundant multi axis vibration table works under load, the coupling between the translation direction and the rotation direction will exist because the centroid of the specimen in the vertical direction is above the control point. Taking the coupling of the X direction and the Ry direction as an example, the acceleration frequency response characteristics between the X direction and the Ry direction are shown in Fig. 5.8 under no-load and full load. As can be seen from the diagram, the coupling between the X and the Ry becomes larger under load. From the frequency response, the coupling under full load is higher than that under the no-load average by 10 dB, thus, it is necessary to adopt decoupling control strategy for the coupling between the X and the Ry. In addition, it can be seen from the diagram that the Ry rotation coupling caused by the translation of the X direction is more than 10 dB higher than the X translation coupling caused by the rotation of the Ry, that is, the coupling of Ry/x is higher than that of x/Ry.

5.4.2 Experiment of Dynamics Decoupling Control Based on Dynamics Model

Using the decoupling control strategy based on dynamic model is proposed, the rotation of Ry direction coupling characteristic caused by the translation of X direction is shown in Fig. 5.9. It can be seen from the figure, without the dynamic decoupling control, the maximum value of Ry/x is 8 dB. The overall decline is about 7 dB using the dynamics decoupling control, and the validity of the

Fig. 5.8 Comparison of coupling characteristic between *x* and *Ry* with and without load

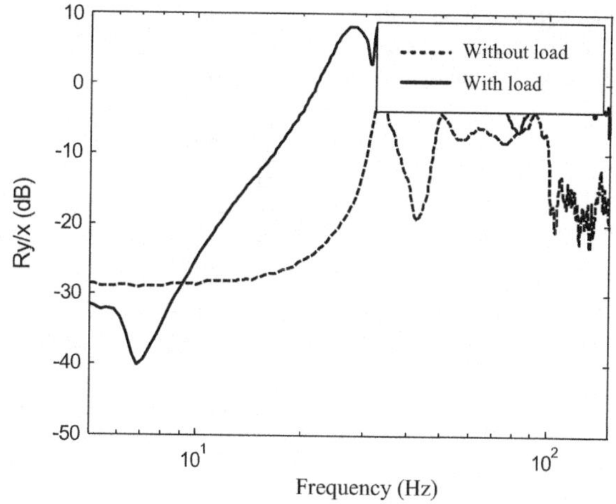

(a) *Ry*/*x*

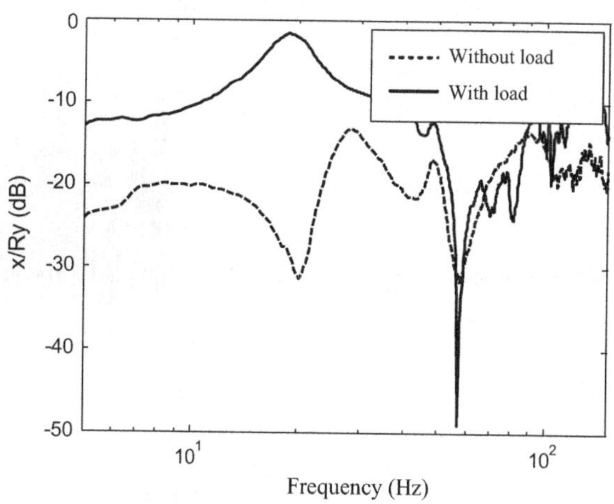

(b) *x*/*Ry*

decoupling control strategy is verified through the time domain waveform of the coupling rotation of Ry direction caused by random acceleration of X direction. Without the dynamic decoupling control, the average amplitude of the time-domain coupled the signal is 4 rad/s^2, the average amplitude drops by half after using dynamics decoupling control.

Using the dynamics decoupling control strategy, the frequency response of the translation of X direction caused by the rotation of Ry direction is shown in Fig. 5.10. From the figure, the decoupling effect of the translation of X direction is poor based on the proposed proportional plus derivative control due to the greater flexibility of base which results in the change of the transfer function between

Fig. 5.9 Coupling in *Ry* direction before and after dynamic decoupling control

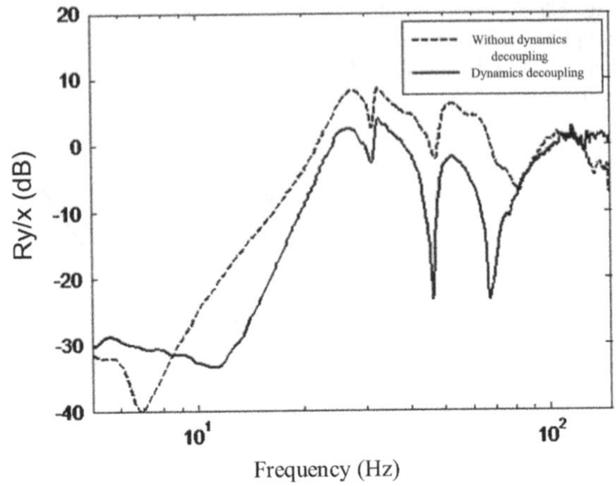

(a) Coupling comparison of *Ry* in frequency domain

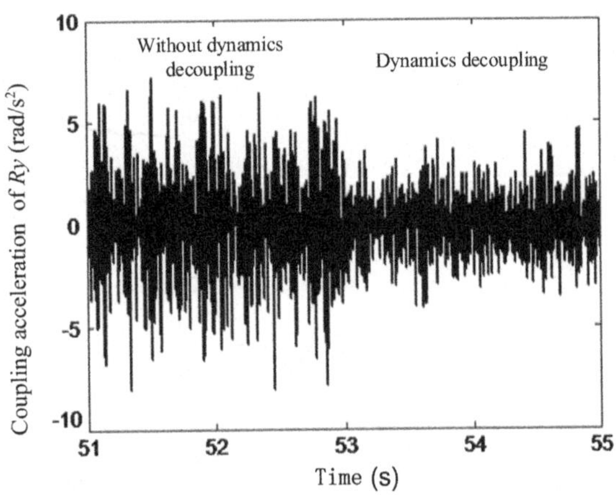

(b) Coupling comparison of *Ry* in time domain

interference force and absolute displacement. In this case, proportional plus derivative control is not effective.

Considering the flexibility of base, the diagram of the feed forward compensation for interference force will be transformed to Fig. 5.11 from Fig. 5.3.

The transfer function between absolute displacement and interference force, the given voltage will be given by

$$L = \frac{k_{sh}\left(\frac{s^2}{\omega_b^2} + \frac{2\varsigma_b}{\omega_b}s + 1\right)}{N(s)}U - \frac{s + \frac{k_{ce}}{A^2}\left(\frac{V_t}{4\beta_e k_{ce}}s + 1\right)\left(\frac{s^2}{\omega_b^2} + \frac{2\varsigma_b}{\omega_b}s + 1\right)}{N(s)}F_d \qquad (5.65)$$

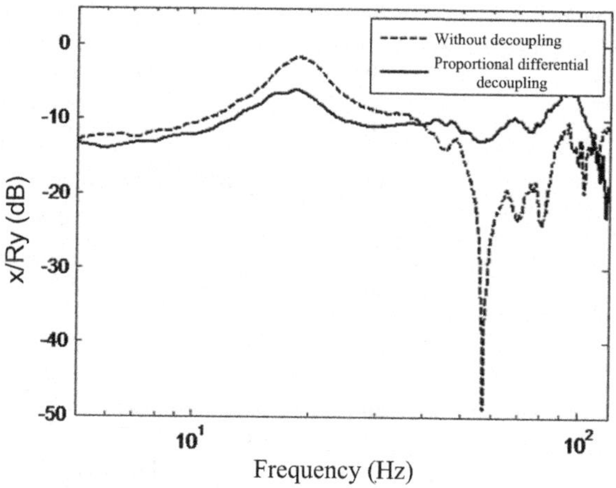

Fig. 5.10 Coupling in x direction before and after dynamic decoupling control

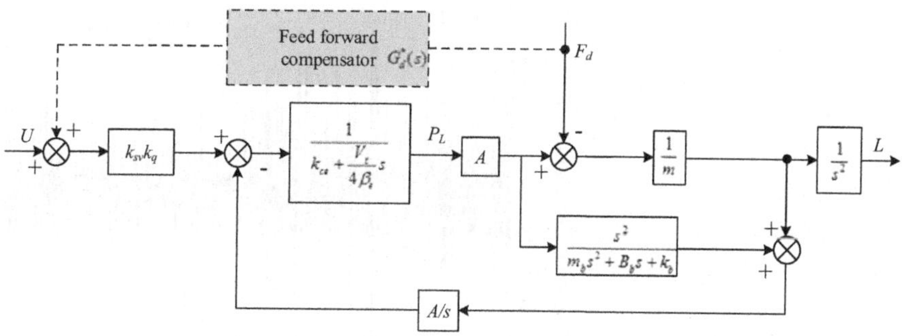

Fig. 5.11 Disturbance force control of hydraulic system under flexible foundation

To make the interference force no effect on the absolute displacement of the payload, the feed forward compensation for interference force will be given by

$$G_d^*(s) = \frac{s}{k_{sh}\left(\frac{s^2}{\omega_b^2} + \frac{2\varsigma_b}{\omega_b}s + 1\right)} + G_d(s) \tag{5.66}$$

From the Eq. (5.66), the improved forward fed compensator is more than the original one, which mainly compensates for the flexible foundation displacement caused by the disturbance force.

Using the improved forward fed compensator, the frequency response of the translation of X direction caused by the rotation of Ry direction is shown in Fig. 5.12.

From the Fig. 5.12, the coupling of X direction reduces significantly using the improved dynamic decoupling control. It can be seen from the figure, without the

Fig. 5.12 Coupling of x direction before and after improved dynamic decoupling control

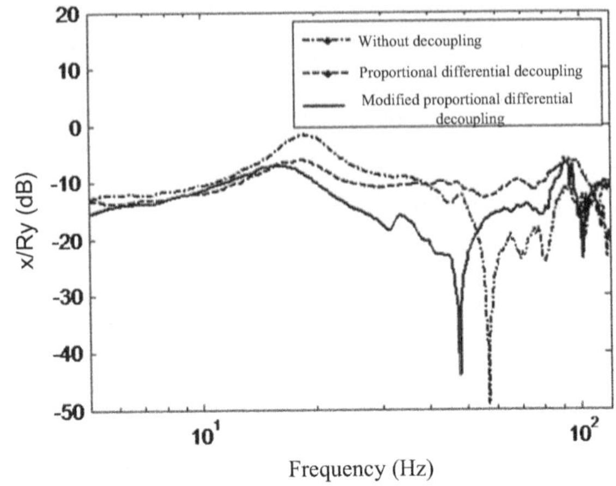

(a) Coupling comparison of x in frequency domain

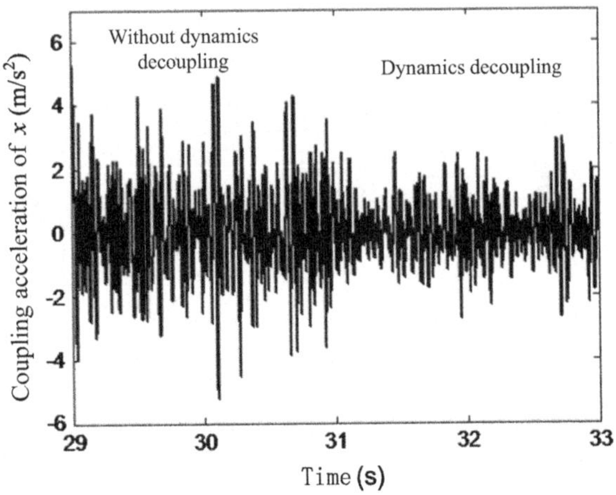

(b) Coupling comparison of x in time domain

dynamic decoupling control, the maximum value of x/Ry is −0.4 dB. The overall decline is about 10 dB using the dynamics decoupling control. The time domain waveform of the coupling of X direction caused by random acceleration of Ry direction also reduces significantly using the decoupling control based on dynamics model.

5.4.3 Experiment of Modal Decoupling Control

The key of modal space decoupling control is how to determine the modal matrix. With respect to the redundant multi-axis vibrating table, it is not applicable to determine the modal matrix correctly through the independent control parameters between the channels, since the control scheme based on linearization of degree of freedom. Not using the decoupling control, the forward gain coefficients of Ry are multiplied by 0.6, 0.8 and 1, respectively, then, the closed-loop frequency response of X direction and Ry direction are shown in Fig. 5.13a, b.

From the Fig. 5.13, the change of control parameters have little effect on the closed-loop frequency response. But, dynamics coupling still exists between x and Ry, thus, it is not concluded that the system is in a modal decoupling state.

The modal matrix is determined by the experimental preceding method. The coupling between the two degrees of freedom near the 20 Hz is relatively large, thus the sinusoidal signal of 20 Hz is used to excite the modal freedom. Elements in modal matrix φ_1 and φ_2 are −0.32 and 0.165, respectively. Take φ_1 for −0.1, −0.32

Fig. 5.13 Closed-loop characteristics of coupling DOF under different control parameters

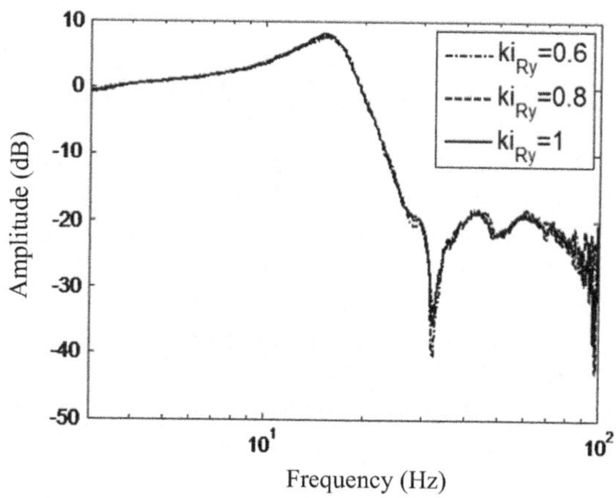

(a) Effect of ki_{Ry} on closed-loop frequency response of x.

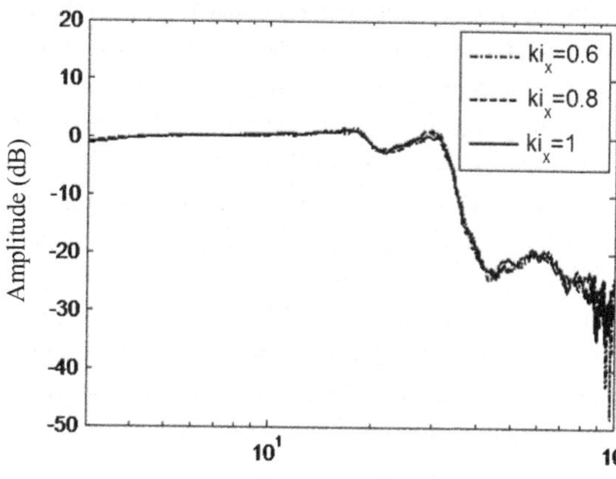

(b) Effect of ki_{Ry} on closed-loop frequency response of Ry.

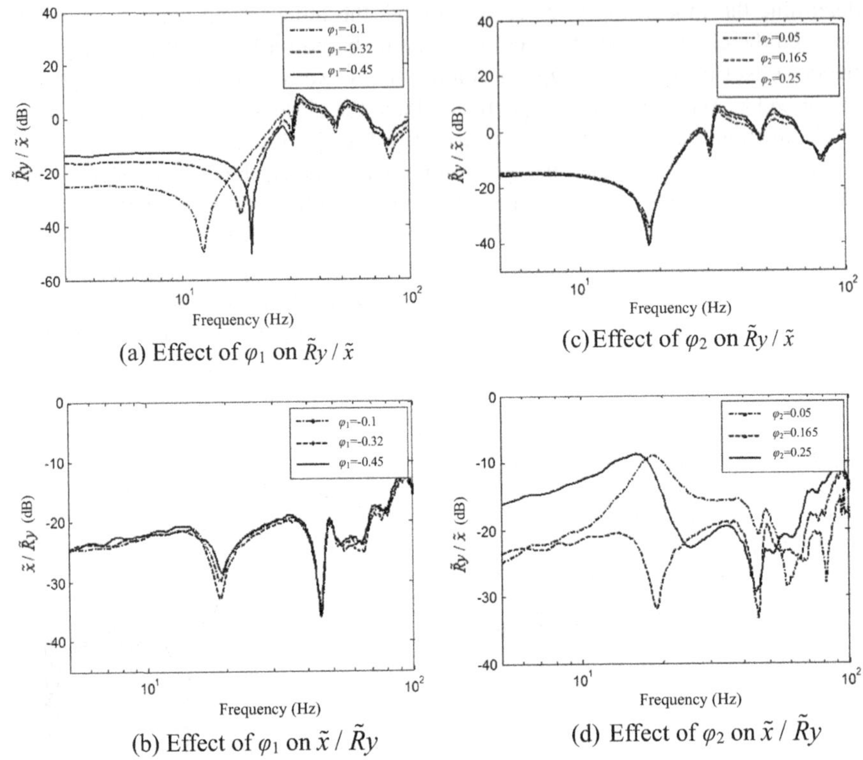

Fig. 5.14 Coupling characteristics under different parameters in modal matrix

and −0.45 respectively, frequency response of $\tilde{R}y/\tilde{x}$ and $\tilde{x}/\tilde{R}y$ are shown in Fig. 5.14a, b.

Changes in φ_1 values affects only the coupling output of $\tilde{R}y$ caused by $\tilde{x}$, and has little effect on $\tilde{x}/\tilde{R}y$. The effect of changes in φ_2 values on $\tilde{R}y/\tilde{x}$ and $\tilde{x}/\tilde{R}y$ is opposite to that in φ_1, the frequency response are shown in Fig. 5.14c, d. It is noteworthy that there is the coupling of X direction caused by flexibility of base, which results in the mode between x and Ry more than two and the coupling between $\tilde{x}$ and $\tilde{R}y$ still exists. Since the modal matrix is determined by 20 Hz sine, the effect of modal space decoupling control is mainly illustrated at low frequency.

To achieve the decoupling control between *x* and *Ry*, the closed-loop frequency response of $\tilde{x}$ and $\tilde{R}y$ should be consistent as much as possible apart from determining the modal matrix. Series feed forward control is performed respectively before $\tilde{x}$ and $\tilde{R}y$, and their dynamic characteristics are adjusted to to be consistent in low frequency domain. The modified amplitude frequency and phase frequency

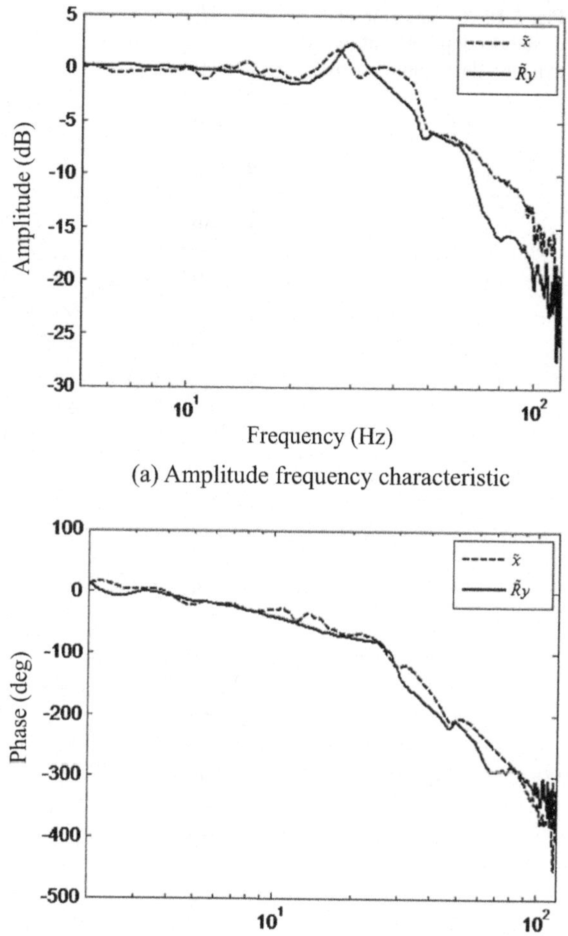

(a) Amplitude frequency characteristic

(b) Phase frequency characteristic

Fig. 5.15 Close-loop characteristics of $\tilde{x}$ and $\tilde{R}y$

responses of $\tilde{x}$ and $\tilde{R}y$ are shown in Fig. 5.15, and they are consistent in low frequency domain.

The acceleration signals are given in x and Ry directions respectively, their coupling frequency response of modal control scheme are shown in Fig. 5.16. From the figure, coupling of x/Ry and Ry/x based on modal decoupling control in low frequency domain is lower than that based on dynamics decoupling control. Coupling of x/Ry and Ry/x decrease by 5 and 10 dB respectively, and the effect of decoupling is significant. The effect of decoupling is not significant due to another mode in high frequency domain.

Fig. 5.16 Comparison of modal decoupling control and dynamic decoupling control

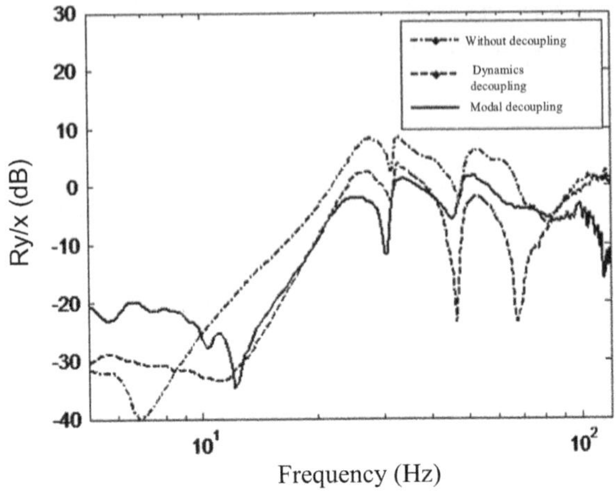

(a) *Ry/x*

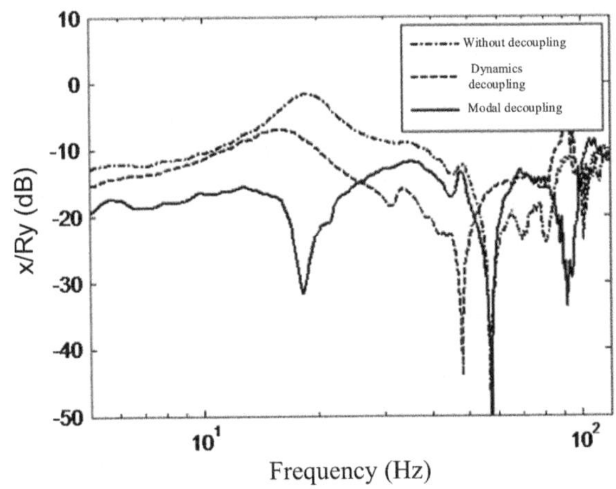

(b) *x/Ry*

5.5 Summary

Firstly, the dynamic model of redundant multi-axial vibration table with load is established, and its dynamic coupling characteristics are analyzed. Based on the model of dynamic coupling, coupling force observer is given. Through the Jacobi matrix, the coupling force is decomposed into the disturbance force of single cylinder, and the decoupling control strategy based on dynamic model is proposed on the basis of studying the feed forward compensation of disturbance force. Then, based on the analysis of the free vibration equation of redundant multi axis shaking table, the transformation from the multi input multiple output system to multiple single input single output systems is achieved through the transformation from the degree of freedom space to the modal space, and the modal decoupling control strategy proposed.

Finally, the method of determining the modal matrix is given by analyzing the relation between the modal matrix and the coupling characteristic of the modal freedom.

Experimental results of dynamic decoupling control indicate that decoupling control strategy based on dynamic model can reduce the effect of rotational coupling due to translation remarkably, but the effect of translation coupling due to rotation is poor. The reason why the effect is poor is the flexibility of the foundation. After the compensation for the flexibility of the foundation, an improved feed forward controller for disturbance force is proposed, which makes the translation coupling caused by rotation reduce obviously.

Firstly, experimental results of modal decoupling control verify the validity of the modal matrix test method. Then, the modal matrix is used to verify the modal space decoupling control. Experimental results indicate that the modal space decoupling control can reduce the coupling between degrees of freedom more remarkably compared to dynamic decoupling control.

Acknowledgements The authors want to acknowledge the support received from Science Challege Project, No. JCKY2016212A506-0107.

References

1. Glazunov V (2010) Design of decoupled parallel manipulators by means of the theory of screws. Mech Mach Theory 45(2):239–250
2. Aminzadeh M, Mahmoodi A, Menhaj MB (2001) A novel decoupling controller design for parallel motion platforms. In: Proceedings of the 8th IEEE international conference on control and automation, Xiamen, China, pp 2086–2091
3. Briot S, Arakelian V, Guegan SP (2009) A new family of partially decoupled parallel manipulator. Mech Mach Theory 44(2):425–444
4. Jin Y, Chen IM (2006) Effects of constraint errors on parallel manipulators with decoupled motion. Mech Mach Theory 41(8):912–928
5. Jin Y, Chen IM, Yang GL (2009) Kinematic design of 6-DOF partially decoupled parallel manipulators. Mech Mach Theory 44(5):912–922
6. Altuzarra O, Loizaga M, Pinto C et al (2010) Synthesis of partially decoupled multi-level manipulators with lower mobility. Mech Mach Theory 45(1):106–118
7. Bristol EH (1966) On a new measure of interaction for multivariable process control. IEEE Trans Autom Control 11(1):133–143
8. Manousiouthakis V, Savage R, Arkum Y (1986) Synthesis of decentralized process control structures using the concept of block relative gain. AIChE J 32(6):991–1003
9. Shen Y, Xie K (1998) The analysis for dynamic measure of structural interaction and its application. J Taiyuan Univ Tech 29(3):233–236
10. Zhang L, Wang J, Wang L (2003) Analysis and simplification of the rigid body dynamic model for a 6-UPS parallel kinematic machine under the uniform motion condition. In: IEEE international conference on robotics, intelligent systems and signal processing, pp 980–985

11. Han J (1996) Development of 3 dimension and 6 degrees of freedom large scale earthquake simulator with three to six degrees of freedom. Harbin, Postdoctoral report of Harbin Institute of Technology, pp 1–65
12. Yang Z, Li H, Han J (2005) Linear model and inertia force decoupling of 3 -dimension hydraulic shaking system. Mach Tool Hydraul, pp 69–71
13. Gizatullin AO, Edge KA (2006) Adaptive control for a multi-axis hydraulic test rig. Proc Inst Mech Eng Part I: J Syst Control Eng 221(2):183–198
14. Ginsberg J, Seemann W (2001) Mechanical and structural vibration: theory and applications. Appl Mech Rev 54(4):60
15. Plummer AR, Guinzio PS (2009) Modal control of an electro hydrostatic flight simulator motion system. In: Proceedings of the ASME dynamic systems and control conference, Hollywood, pp 1257–1264
16. Plummer AR (2007) High bandwidth motion control for multi-axis servo hydraulic mechanisms. In: Proceedings of 2007 ASME international mechanical engineering congress and exposition, Seattle, 41240
17. Ogbobe Peter O, ZhengMao Y, Jiang H et al (2010) Modal space decoupled controller for hydraulically driven six degree of freedom parallel robot. In: Proceedings of 2010 2nd international conference on mechanical and electronics engineering, Kyto, Japan, pp 1280–1284
18. Yang C, Huang Q, Han J (2012) Decoupling control for spatial six-degree-of-freedom electro-hydraulic parallel robot. Robot Comput Integr Manufact 28(1):14–23
19. Yang C, Han J (2013) Dynamic coupling analysis of a spatial 6-DOF electro-hydraulic parallel manipulator using a modal decoupling method. Int J Adv Rob Syst 10(104):1–8
20. Jiang H, He J, Tong Z (2012) Modal space control for a hydraulically driven stewart platform. J Control Eng Tech 2(3):106–115
21. Yang C, Han J (2013) Decoupled-space control and experimental evaluation of spatial electro-hydraulic robotic manipulators using SVD algorithms. IEEE Trans Ind Electron 61 (7):3427–3438
22. Staicu S (2011) Dynamics of the 6-6 stewart parallel manipulator. Robot Comput Integr Manufact 27(1):212–220
23. Yang C, Huang Q, Jiang H et al (2010) PD control with gravity compensation for hydraulic 6-DOF parallel manipulator. Mech Mach Theory 45(4), 666–677

Chapter 6
Design and Analysis of the 6-DOF Decoupled Parallel Kinematics Mechanism

Victor Glazunov, Natalya Nosova, Sergey Kheylo and Andrey Tsarkov

Abstract Parallel mechanisms (PMs) with six degree of freedoms (DOF) are widely used in such different segments of industry as a measuring, tooling, and positioning device. The spatial 6-DOF mechanism is examined in this chapter. The mechanism studied has both kinematic and dynamic decoupling. The kinematic problem solution is presented. The control algorithm of handling PMs was tested on the basis of nonlinear systems control theory. The velocity and control problems were solved with the use of dynamic and kinematic decoupling.

6.1 Introduction

The increase of DOF number leads to the capacity improvement of parallel structure mechanisms (PMs) but makes the solution of kinematic, dynamic, and control problems more complicated. To simplify the task solution, the mechanisms with kinematic and dynamic decoupled motions are considered to be more preferable. Therefore, decoupler application permits to analyze 4-DOF, 5-DOF, and 6-DOF mechanisms as two separate ones, which perform rotational and translational motions.

V. Glazunov · N. Nosova
The Blagonravov Mechanical Engineering Research Institute of the Russian Academy of Sciences, Moscow, Russia

S. Kheylo (✉)
The Kosygin State University of Russia, Moscow, Russia
e-mail: sheilo@yandex.ru

A. Tsarkov
Bauman Moscow State Technical University, Kaluga Branch, Russia

V. Arakelian (ed.), *Dynamic Decoupling of Robot Manipulators*, Mechanisms and Machine Science 56, https://doi.org/10.1007/978-3-319-74363-9_6

6.2 Structural Synthesis and Analysis of 4-DOF, 5-DOF, and 6-DOF Parallel Structure Mechanisms

The development of well-known robot ORTHOGLIDE [1] is considered in this chapter. It involves adding the rotational motion. Structural synthesis and analysis of the developed structures are performed. The operability of the mechanisms obtained, critical positions and incontrollable mobility are determined by means of a screw method. The problems considered in this chapter are necessary for further studies of mechanisms.

6.2.1 Schematic Structure of 4-DOF Parallel Mechanism

Let us consider the mechanism which is the development of ORTHOGLIDE robot with one rotational added DOF, with both rotation and prismatic motion being provided by means of the same kinematic chain (Fig. 6.1) [2].

Determine the DOF number of the manipulator using the Malyshev formula for spatial mechanisms

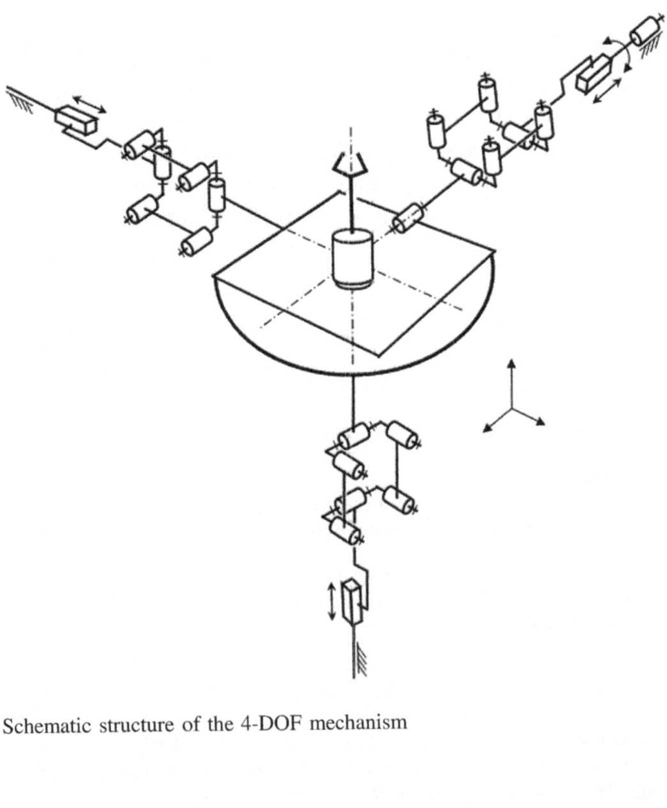

Fig. 6.1 Schematic structure of the 4-DOF mechanism

$$W = 6 \cdot (n - 1) - 5 \cdot p_5 - 4 \cdot p_4$$

where n is the number of the links;
p_5 is the number of the fifth class pairs (one-DOF kinematic pairs)
p_4 is the number of the fourth class pairs (two-DOF kinematic pairs)
Parallel link mechanism, being a plain mechanism, possesses one DOF so it can be regarded as one-degree-of-freedom prismatic kinematic pair. Revolute parallel link mechanism has one degree of freedom according to the Chebychev formula

$$W = 3 \cdot (n - 1) - 2 \cdot p_5 - p_4$$

where n is the number of links;
p_5 is the number of the fifth class pairs (one-DOF kinematic pairs);
p_4 is the number of the fourth class pairs (two-DOF kinematic pairs);
Therefore, we get

$$W = 3 \cdot (4 - 1) - 2 \cdot 4 = 1$$

Let us consider the DOF number under this condition. Twelve rotating pairs of three parallel link mechanisms are replaced by three prismatic kinematic pairs.

Let us substitute revolute joint parallel mechanisms for universal joint in two kinematic chains, which don't possess the rotational movement driver. The third kinematic chain contains the rotational movement driver. Suppose the third chain revolute joint parallel mechanism is a prismatic kinematic pair (Fig. 6.2) [3].

In this case we have 15 links and 16 one-DOF kinematic pairs. Having substituted these values into the formula above, obtain

$$W = 6 \cdot (15 - 1) - 5 \cdot 16 = 4$$

Thus, the number of freedom is equal to 4.

To determine the working capacity of given mechanisms the Plucker coordinates of kinematic pair axes unit vectors should be considered. This method permits to identify potential critical positions (singularities) as well as incontrollable mobility.

Here for every kinematic pair there is a unit vector of the pair's axis and the Plucker coordinate matrix for the unit vectors is made up. Corresponding composition is shown on the Fig. 6.3.

Let us consider the Plucker coordinates for the unit screws of the kinematic pairs axes. The mechanism is in the origin position (Fig. 6.3). The unit screws, which specify the position of kinematic pairs axes, possess the following coordinates: $\boldsymbol{E}_{11}$ (1, 0, 0, 0, 0, 0), $\boldsymbol{E}_{12}$ (0, 0, 0, 1, 0, 0), $\boldsymbol{E}_{13}$ (0, −1, 0, 0, 0, e^0_{13z}), E_{14} (0, 0, 0, 0, 1, 0), $\boldsymbol{E}_{15}$ (0, −1, 0, 0, 0, e^0_{15z}), $\boldsymbol{E}_{16}$ (1, 0, 0, 0, 0, 0); $\boldsymbol{E}_{22}$ (0, 0, 0, 0, 1, 0), $\boldsymbol{E}_{23}$ (0, 0, 1, e^0_{23x}, 0, 0), $\boldsymbol{E}_{24}$ (0, 0, 0, 0, 0, −1), $\boldsymbol{E}_{25}$ (0, 0, 1, e^0_{25x}, 0, 0); $\boldsymbol{E}_{26}$ (0, 0, 0, 0, 0, 1), $\boldsymbol{E}_{33}$ (−1, 0, 0, 0, e^0_{33y}, 0), $\boldsymbol{E}_{34}$ (0, 0, 0, 1, 0, 0), $\boldsymbol{E}_{35}$ (−1, 0, 0, 0, e^0_{35y}, 0).

We shall deal with the first kinematic chain. Since the unit vectors $\boldsymbol{E}_{11}$ and $\boldsymbol{E}_{16}$ are in the same axis, take no notice of unit vector $\boldsymbol{E}_{16}$. In this case we have five unit

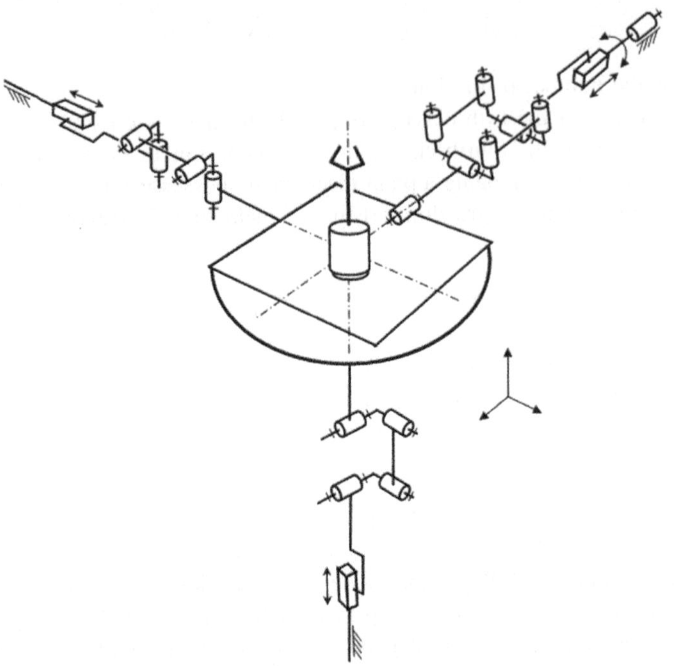

Fig. 6.2 Schematic structure of the 4-DOF mechanism with substitute revolute joint parallel mechanism for the universal joint

vectors. Reciprocity conditions for the Plucker coordinates of unit screws and wrench may be expressed as five equations out of which it is necessary to find one indeterminate and consequently get one wrench:

$$r_x \cdot e^0_{11x} + r_y \cdot e^0_{11y} + r_z \cdot e^0_{11z} + r^0_x \cdot e_{11x} + r^0_y \cdot e_{11y} + r^0_z \cdot e_{11z} = 0$$
$$r_x \cdot 0 + r_y \cdot 0 + r_z \cdot 0 + r^0_x \cdot 1 + r^0_y \cdot 0 + r^0_z \cdot 0 = 0$$
$$r_x \cdot e^0_{12x} + r_y \cdot e^0_{12y} + r_z \cdot e^0_{12z} + r^0_x \cdot e_{12x} + r^0_y \cdot e_{12y} + r^0_z \cdot e_{12z} = 0$$
$$r_x \cdot 1 + r_y \cdot 0 + r_z \cdot 0 + r^0_x \cdot 0 + r^0_y \cdot 0 + r^0_z \cdot 0 = 0$$
$$r_x \cdot e^0_{13x} + r_y \cdot e^0_{13y} + r_z \cdot e^0_{13z} + r^0_x \cdot e_{13x} + r^0_y \cdot e_{13y} + r^0_z \cdot e_{13z} = 0$$
$$r_x \cdot 0 + r_y \cdot 0 + r_z \cdot e^0_{13z} + r^0_x \cdot 0 + r^0_y \cdot (-1) + r^0_z \cdot 0 = 0$$
$$r_x \cdot e^0_{14x} + r_y \cdot e^0_{14y} + r_z \cdot e^0_{14z} + r^0_x \cdot e_{14x} + r^0_y \cdot e_{14y} + r^0_z \cdot e_{14z} = 0$$
$$r_x \cdot 0 + r_y \cdot 1 + r_z \cdot 0 + r^0_x \cdot 0 + r^0_y \cdot 0 + r^0_z \cdot 0 = 0$$
$$r_x \cdot e^0_{15x} + r_y \cdot e^0_{15y} + r_z \cdot e^0_{15z} + r^0_x \cdot e_{15x} + r^0_y \cdot e_{15y} + r^0_z \cdot e_{15z} = 0$$
$$r_x \cdot 0 + r_y \cdot 0 + r_z \cdot e^0_{15z} + r^0_x \cdot 0 + r^0_y \cdot (-1) + r^0_z \cdot 0 = 0$$

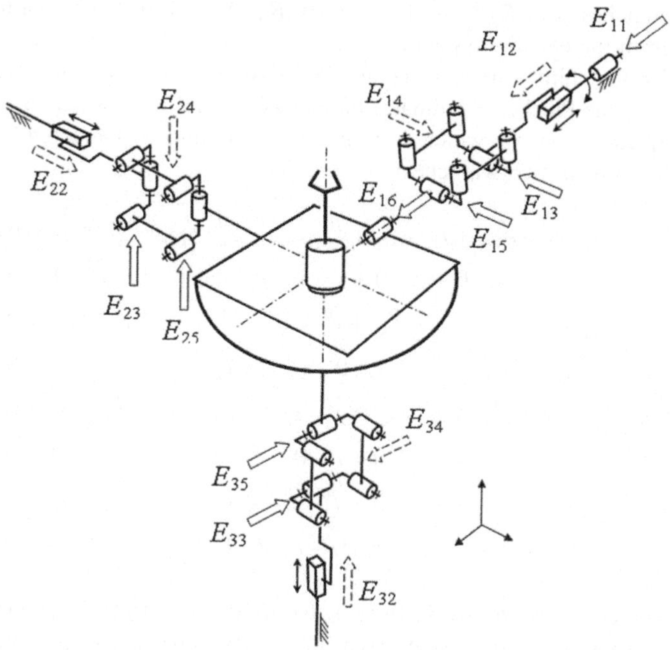

Fig. 6.3 Schematic structure of the 4-DOF parallel mechanism

where e_{ijx}, e_{ijy}, e_{ijz}, e^0_{ijx}, e^0_{ijy}, e^0_{ijz} are the vector and moment parts of the unit screw, respectively (i = 1–3; j = 1–3); r_{ijx}, r_{ijy}, r_{ijz}, r^0_{ijx}, r^0_{ijy}, r^0_{ijz} are the vector and moment parts of the power screw, accordingly (i = 1–3; j = 1–3).

One wrench with $\boldsymbol{R}_1$ (0, 0, 0, 0, 0, 1) coordinates is obtained from the equation system.

Consider the second kinematic chain. We have four unit vectors. Reciprocity conditions for the Plucker coordinates of unit screws and the wrench may be expressed as four equations out of which it is necessary to find two indeterminates and consequently obtain two wrenches:

$$r_x \cdot e^0_{22x} + r_y \cdot e^0_{22y} + r_z \cdot e^0_{22z} + r^0_x \cdot e_{22x} + r^0_y \cdot e_{22y} + r^0_z \cdot e_{22z} = 0$$
$$r_x \cdot 0 + r_y \cdot 1 + r_z \cdot 0 + r^0_x \cdot 0 + r^0_y \cdot 0 + r^0_z \cdot 0 = 0$$
$$r_x \cdot e^0_{23x} + r_y \cdot e^0_{23y} + r_z \cdot e^0_{23z} + r^0_x \cdot e_{23x} + r^0_y \cdot e_{23y} + r^0_z \cdot e_{23z} = 0$$
$$r_x \cdot e^0_{23x} + r_y \cdot 0 + r_z \cdot 0 + r^0_x \cdot 0 + r^0_y \cdot 0 + r^0_z \cdot 1 = 0$$
$$r_x \cdot e^0_{24x} + r_y \cdot e^0_{24y} + r_z \cdot e^0_{24z} + r^0_x \cdot e_{24x} + r^0_y \cdot e_{24y} + r^0_z \cdot e_{24z} = 0$$
$$r_x \cdot 0 + r_y \cdot 0 + r_z \cdot (-1) + r^0_x \cdot 0 + r^0_y \cdot 0 + r^0_z \cdot 0 = 0$$
$$r_x \cdot e^0_{25x} + r_y \cdot e^0_{25y} + r_z \cdot e^0_{25z} + r^0_x \cdot e_{25x} + r^0_y \cdot e_{25y} + r^0_z \cdot e_{25z} = 0$$
$$r_x \cdot e^0_{25x} + r_y \cdot 0 + r_z \cdot 0 + r^0_x \cdot 0 + r^0_y \cdot 0 + r^0_z \cdot 1 = 0$$

Two wrenches with $\boldsymbol{R}_2$ (0, 0, 0, 1, 0, 0), $\boldsymbol{R}_3$ (0, 0, 0, 0, 1, 0) coordinates are obtained from the equation system.

Consider the third kinematic chain. We have four unit vectors. Reciprocity conditions for the Plucker coordinates of unit screws and wrench may be expressed as four equations out of which it is necessary to find two indeterminates and consequently obtain two wrenches:

$$
\begin{aligned}
r_x \cdot e^0_{32x} + r_y \cdot e^0_{32y} + r_z \cdot e^0_{32z} + r^0_x \cdot e_{32x} + r^0_y \cdot e_{32y} + r^0_z \cdot e_{32z} &= 0 \\
r_x \cdot 0 + r_y \cdot 0 + r_z \cdot 1 + r^0_x \cdot 0 + r^0_y \cdot 0 + r^0_z \cdot 0 &= 0 \\
r_x \cdot e^0_{33x} + r_y \cdot e^0_{33y} + r_z \cdot e^0_{33z} + r^0_x \cdot e_{33x} + r^0_y \cdot e_{33y} + r^0_z \cdot e_{33z} &= 0 \\
r_x \cdot 0 + r_y \cdot e^0_{33y} + r_z \cdot 0 + r^0_x \cdot (-1) + r^0_y \cdot 0 + r^0_z \cdot 0 &= 0 \\
r_x \cdot e^0_{34x} + r_y \cdot e^0_{34y} + r_z \cdot e^0_{34z} + r^0_x \cdot e_{34x} + r^0_y \cdot e_{34y} + r^0_z \cdot e_{34z} &= 0 \\
r_x \cdot 1 + r_y \cdot 0 + r_z \cdot 0 + r^0_x \cdot 0 + r^0_y \cdot 0 + r^0_z \cdot 0 &= 0 \\
r_x \cdot e^0_{35x} + r_y \cdot e^0_{35y} + r_z \cdot e^0_{35z} + r^0_x \cdot e_{35x} + r^0_y \cdot e_{35y} + r^0_z \cdot e_{35z} &= 0 \\
r_x \cdot 0 + r_y \cdot e^0_{35y} + r_z \cdot 0 + r^0_x \cdot (-1) + r^0_y \cdot 0 + r^0_z \cdot 0 &= 0
\end{aligned}
$$

Two wrenches with $\boldsymbol{R}_4$ (0, 0, 0, 0, 1, 0), $\boldsymbol{R}_5$ (0, 0, 0, 0, 0, 1) coordinates are obtained from the equation system wrench coordinates which can be represented as a corresponding matrix $\boldsymbol{R}$ (6 × 5)

$$
\boldsymbol{R} = \begin{pmatrix} 0 & 0 & 0 & 0 & 0 & 1 \\ 0 & 0 & 0 & 1 & 0 & 0 \\ 0 & 0 & 0 & 0 & 1 & 0 \\ 0 & 0 & 0 & 0 & 1 & 0 \\ 0 & 0 & 0 & 0 & 0 & 1 \end{pmatrix}
$$

Only three independent wrenches can be obtained from this matrix. These are the screws of infinity pitch (moments) of the infinity great characteristic quantity (moments) constraining the rotation.

Consider separately the first kinematic chain, the latter is being translated about an axis of reference, and determine the chain Plucker coordinates.

Consider the Plucker coordinates of the unit screws of mechanism kinematic pair axes (Fig. 6.4). The coordinates of the unit screws, which determine the kinematic pair axes location, are the same: $\boldsymbol{E}_{11}$ (1, 0, 0, 0, 0, 0), $\boldsymbol{E}_{12}$ (0, 0, 0, 1, 0, 0), $\boldsymbol{E}_{13}$ (0, −1, 0, 0, 0, e^0_{13z}), $\boldsymbol{E}_{14}$ (0, 0, 0, e^0_{14x}, e^0_{14y}, 0), $\boldsymbol{E}_{15}$ (0, −1, 0, 0, 0, e^0_{15z}), $\boldsymbol{E}_{16}$ (1, 0, 0, 0, 0, e^0_{16z}).

Set up the matrix from the obtained unit screws coordinates:

$$\boldsymbol{E} = \begin{pmatrix} 1 & 0 & 0 & 0 & 0 & 0 \\ 0 & 0 & 0 & 1 & 0 & 0 \\ 0 & -1 & 0 & 0 & 0 & e^0_{13z} \\ 0 & 0 & 0 & e^0_{14x} & e^0_{15y} & 0 \\ 0 & -1 & 0 & 0 & 0 & e^0_{15z} \\ 1 & 0 & 0 & 0 & 0 & e^0_{16z} \end{pmatrix}$$

Remove zero column from the obtained matrix:

$$\boldsymbol{E}' = \begin{pmatrix} 1 & 0 & 0 & 0 & 0 \\ 0 & 0 & 1 & 0 & 0 \\ 0 & -1 & 0 & 0 & e^0_{13z} \\ 0 & 0 & e^0_{14x} & e^0_{14y} & 0 \\ 0 & -1 & 0 & 0 & e^0_{15z} \\ 1 & 0 & 0 & 0 & e^0_{16z} \end{pmatrix}$$

Write out 5 × 5 matrixes deleting one matrix row by another.

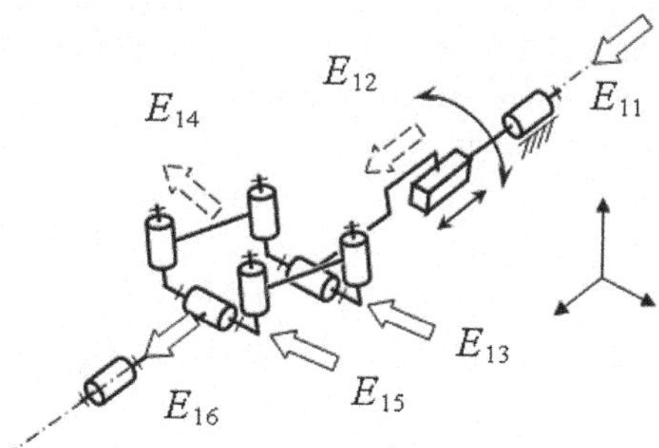

Fig. 6.4 First kinematic chain of the 4-DOF mechanism

$$E_1 = \begin{pmatrix} 0 & 0 & 1 & 0 & 0 \\ 0 & -1 & 0 & 0 & e^0_{13z} \\ 0 & 0 & e^0_{14x} & e^0_{14y} & 0 \\ 0 & -1 & 0 & 0 & e^0_{15z} \\ 1 & 0 & 0 & 0 & e^0_{16z} \end{pmatrix}; \; E_2 = \begin{pmatrix} 1 & 0 & 0 & 0 & 0 \\ 0 & -1 & 0 & 0 & e^0_{13z} \\ 0 & 0 & e^0_{14x} & e^0_{14y} & 0 \\ 0 & -1 & 0 & 0 & e^0_{15z} \\ 1 & 0 & 0 & 0 & e^0_{16z} \end{pmatrix}$$

$$E_3 = \begin{pmatrix} 1 & 0 & 0 & 0 & 0 \\ 0 & 0 & 1 & 0 & 0 \\ 0 & 0 & e^0_{14x} & e^0_{14y} & 0 \\ 0 & -1 & 0 & 0 & e^0_{15z} \\ 1 & 0 & 0 & 0 & e^0_{16z} \end{pmatrix}; \; E_4 = \begin{pmatrix} 1 & 0 & 0 & 0 & 0 \\ 0 & 0 & 1 & 0 & 0 \\ 0 & -1 & 0 & 0 & e^0_{13x} \\ 0 & -1 & 0 & 0 & e^0_{15z} \\ 1 & 0 & 0 & 0 & e^0_{16z} \end{pmatrix};$$

$$E_5 = \begin{pmatrix} 1 & 0 & 0 & 0 & 0 \\ 0 & 0 & 1 & 0 & 0 \\ 0 & -1 & 0 & 0 & e^0_{13x} \\ 0 & 0 & e^0_{14x} & e^0_{14y} & 0 \\ 1 & 0 & 0 & 0 & e^0_{15z} \end{pmatrix}$$

Let us analyze these matrixes with relation to the possibility of the chain internal mobility. It can be shown that the partial chain containing four kinematic pairs (except the translatory device) possesses the mobility.

Let us consider the matrix $\boldsymbol{E_2}$ trying to express the end matrix row via the first four rows. The problem will reduce to the system of linear equations as regard to the scalar factors. The third row multiplying factor is evident to be equal zero. We get a three equation system with respect to three unknown variables. Write the system down:

$$\begin{aligned} a \cdot 1 + 0 + 0 + 0 &= 1, \\ 0 + b \cdot (-1) + 0 + c \cdot (-1) &= 0, \\ 0 + b \cdot e^0_{13z} + 0 + c \cdot e^0_{15z} &= e^0_{15z} \end{aligned}$$

Solve this system

$$a = 1, \quad b = -c, \quad b \cdot e^0_{13z} = e^0_{16z} - c \cdot e^0_{15z}, c = \frac{e^0_{16z}}{e^0_{15} - e^0_{13z}},$$

Thus stated coefficients determining speeds in the corresponding kinematic pairs are evident to be found. With the internal mobility in this chain it is possible to transmit rotation from the rotation device to the rotation kinematic pair coupled with the output link.

The similar analysis can be done for a modified mechanism which will be possessed by the unit screws of kinematic pairs (Fig. 6.5).

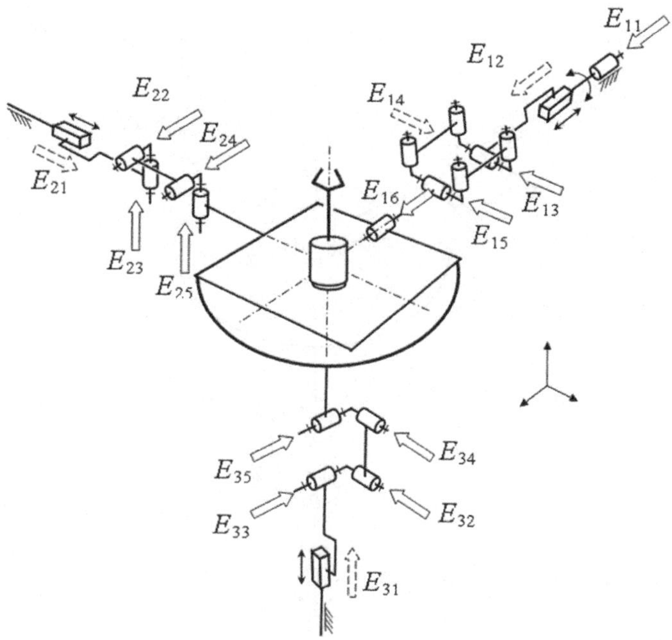

Fig. 6.5 Schematic structure of the 5-DOF parallel mechanism

6.2.2 Schematic Structure of the 5-DOF Parallel Mechanism

Consider the 5-DOF mechanism after having added two rotational motion drivers in two kinematic pairs to the original scheme of the ORTHOGLIDE mechanism (Fig. 6.6) [4].

Twelve rotational kinematic pairs of three parallelograms are substituted for three prismatic kinematic pairs. Substitute the revolute joint parallel mechanism for the universal joint in the kinematic chain, which does not contain the rotational motor driver. Assume that parallel link mechanisms are prismatic kinematic pairs in two other kinematic chains, which contain the rotational motion drivers (Fig. 6.7).

In this case we have 16 links and 17 1-DOF kinematic pairs. DOF number is

$$W = 6 \cdot (16-1) - 5 \cdot 17 = 5$$

To determine working capacity of given mechanisms one should consider the Plucker coordinates of basis vectors of the kinematic pairs axes. This method permits to identify potential critical positions (singularities) as well as uncontrolled mobility.

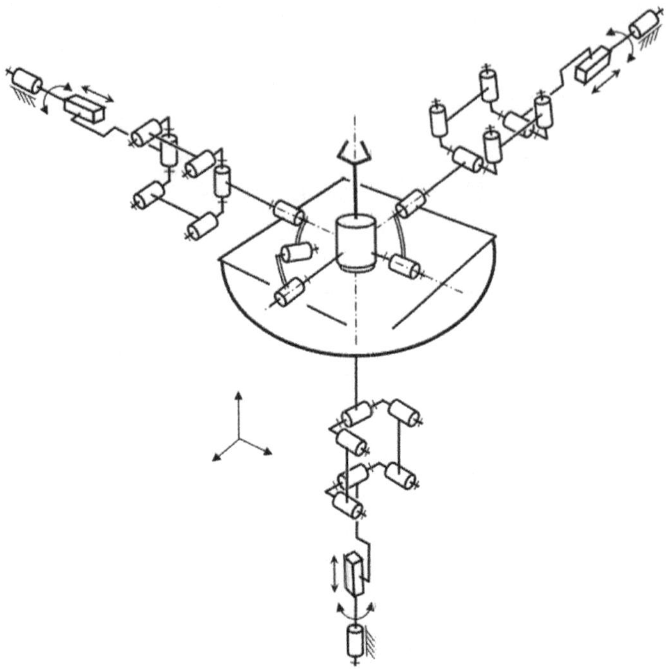

Fig. 6.6 Schematic structure of the 5-DOF mechanism

In this case each kinematic pair matches its axis basis vector and the Pluecker coordinates matrix of these basis vectors is compiled. The corresponding design is on the Fig. 6.8.

Thus, regard the Plucker coordinates for the unit screws of the kinematic pairs axes of mechanism that is in its starting point (Fig. 6.8). The kinematic pairs axes position is determined by units screws with coordinates: $\boldsymbol{E}_{11}$ (1, 0, 0, 0, 0, 0), $\boldsymbol{E}_{12}$ (0, 0, 0, 1, 0, 0), $\boldsymbol{E}_{13}$ (0, −1, 0, 0, 0, e^0_{13z}), $\boldsymbol{E}_{14}$ (0, 0, 0, 0, 1, 0), $\boldsymbol{E}_{15}$ (0, −1, 0, 0, 0, e^0_{15z}), $\boldsymbol{E}_{16}$ (1, 0, 0, 0, 0, 0); $\boldsymbol{E}_{21}$ (0, 1, 0, 0, 0, 0), $\boldsymbol{E}_{22}$ (0, 0, 0, 0, 1, 0), $\boldsymbol{E}_{23}$ (0, 0, 1, e^0_{23x}, 0, 0), $\boldsymbol{E}_{24}$ (0, 0, 0, 0, 0, −1), $\boldsymbol{E}_{25}$ (0, 0, 1, e^0_{25x}, 0, 0), $\boldsymbol{E}_{26}$ (0, 1, 0, 0, 0, 0), $\boldsymbol{E}_{32}$ (0, 0, 0, 0, 0, 1), $\boldsymbol{E}_{33}$ (−1, 0, 0, 0, e^0_{33y}, 0), $\boldsymbol{E}_{34}$ (0, 0, 0, 1, 0, 0), $\boldsymbol{E}_{35}$ (−1, 0, 0, 0, e^0_{35y}, 0).

Examine the first kinematic chain. Since unit vectors $\boldsymbol{E}_1$ and $\boldsymbol{E}_{16}$ are on the same axis, then the unit vector $\boldsymbol{E}_{16}$ is not taken into account. Therefore, we have five unit vectors. Reciprocation conditions for the Plucker coordinates of the unit screws and wrench can be written as five equations. For the solution to be found it is necessary to obtain one unknown and consequently one wrench:

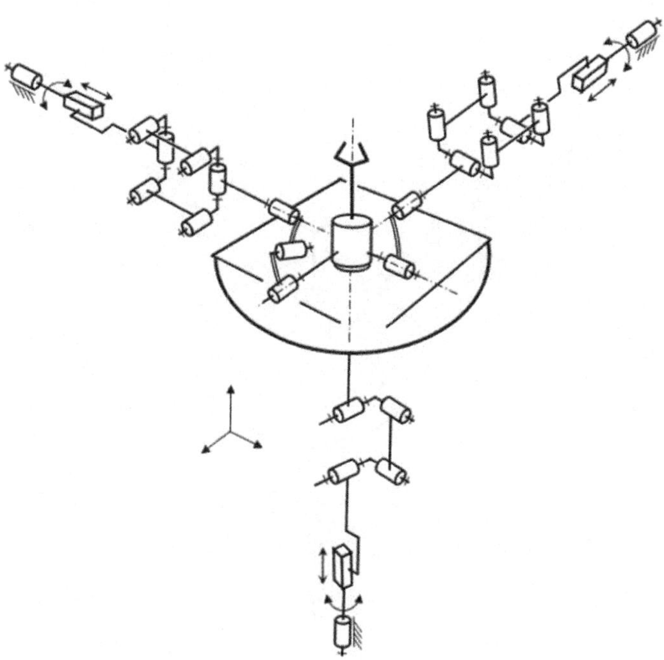

Fig. 6.7 Schematic structure of the 5-DOF mechanism in which revolute joint parallel mechanisms are substituted for the universal joint

$$
\begin{aligned}
r_x \cdot e^0_{11x} + r_y \cdot e^0_{11y} + r_z \cdot e^0_{11z} + r^0_x \cdot e_{11x} + r^0_y \cdot e_{11y} + r^0_z \cdot e_{11z} &= 0 \\
r_x \cdot 0 + r_y \cdot 0 + r_z \cdot 0 + r^0_x \cdot 1 + r^0_y \cdot 0 + r^0_z \cdot 0 &= 0 \\
r_x \cdot e^0_{12x} + r_y \cdot e^0_{12y} + r_z \cdot e^0_{12z} + r^0_x \cdot e_{12x} + r^0_y \cdot e_{12y} + r^0_z \cdot e_{12z} &= 0 \\
r_x \cdot 1 + r_y \cdot 0 + r_z \cdot 0 + r^0_x \cdot 0 + r^0_y \cdot 0 + r^0_z \cdot 0 &= 0 \\
r_x \cdot e^0_{13x} + r_y \cdot e^0_{13y} + r_z \cdot e^0_{13z} + r^0_x \cdot e_{13x} + r^0_y \cdot e_{13y} + r^0_z \cdot e_{13z} &= 0 \\
r_x \cdot 0 + r_y \cdot 0 + r_z \cdot e^0_{13z} + r^0_x \cdot 0 + r^0_y \cdot (-1) + r^0_z \cdot 0 &= 0 \\
r_x \cdot e^0_{14x} + r_y \cdot e^0_{14y} + r_z \cdot e^0_{14z} + r^0_x \cdot e_{14x} + r^0_y \cdot e_{14y} + r^0_z \cdot e_{14z} &= 0 \\
r_x \cdot 0 + r_y \cdot 1 + r_z \cdot 0 + r^0_x \cdot 0 + r^0_y \cdot 0 + r^0_z \cdot 0 &= 0 \\
r_x \cdot e^0_{15x} + r_y \cdot e^0_{15y} + r_z \cdot e^0_{15z} + r^0_x \cdot e_{15x} + r^0_y \cdot e_{15y} + r^0_z \cdot e_{15z} &= 0 \\
r_x \cdot 0 + r_y \cdot 0 + r_z \cdot e^0_{15z} + r^0_x \cdot 0 + r^0_y \cdot (-1) + r^0_z \cdot 0 &= 0
\end{aligned}
$$

where $e_{ijx}, e_{ijy}, e_{ijz}, e^0_{ijx}, e^0_{ijy}, e^0_{ijz}$ are the unit screw vector and moment parts, respectively (i = 1–3; j = 1–3); $r_{ijx}, r_{ijy}, r_{ijz}, r^0_{ijx}, r^0_{ijy}, r^0_{ijz}$ are the wrench vector and moment parts, respectively (i = 1–3; j = 1–3).

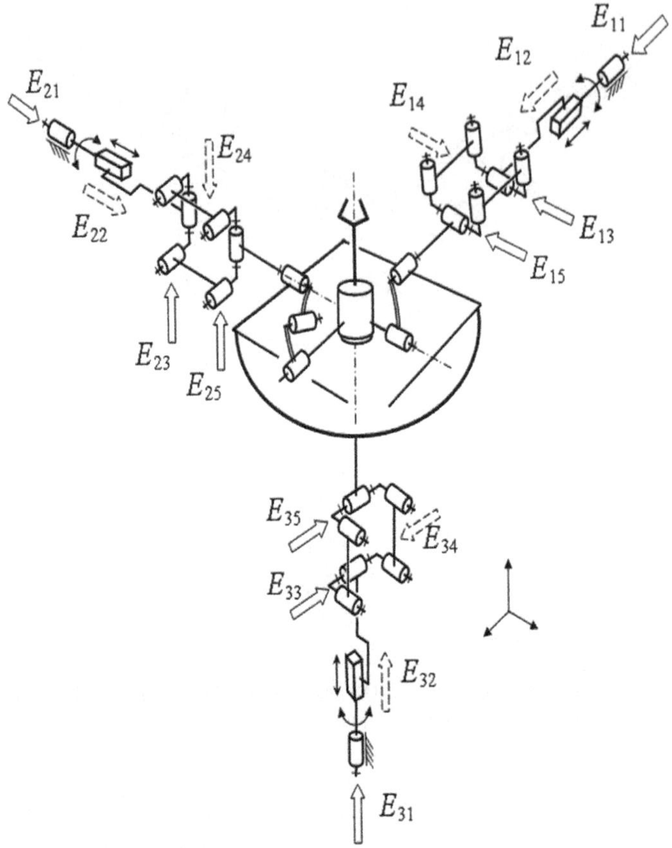

Fig. 6.8 Schematic structure of the 5-DOF parallel mechanism

We can obtain one wrench with $\boldsymbol{R}_1$ (0, 0, 0, 0, 0, 1) coordinates from the system of equations.

Consider the second kinematic chain. Since unit vectors $\boldsymbol{E}_{21}$ and $\boldsymbol{E}_{26}$ are on the same axis, then the unit vector E_{26} is not taken into account. Therefore, we have five unit vectors. Reciprocation conditions for the Plucker coordinates of the unit screws and wrenches can be written as five equations. For the solution to be found it is necessary to obtain one unknown and consequently one wrench:

$$r_x \cdot e^0_{21x} + r_y \cdot e^0_{21y} + r_z \cdot e^0_{21z} + r^0_x \cdot e_{21x} + r^0_y \cdot e_{21y} + r^0_z \cdot e_{21z} = 0$$
$$r_x \cdot 0 + r_y \cdot 0 + r_z \cdot 0 + r^0_x \cdot 0 + r^0_y \cdot 1 + r^0_z \cdot 0 = 0$$
$$r_x \cdot e^0_{22x} + r_y \cdot e^0_{22y} + r_z \cdot e^0_{22z} + r^0_x \cdot e_{22x} + r^0_y \cdot e_{22y} + r^0_z \cdot e_{22z} = 0$$
$$r_x \cdot 0 + r_y \cdot 1 + r_z \cdot 0 + r^0_x \cdot 0 + r^0_y \cdot 0 + r^0_z \cdot 0 = 0$$
$$r_x \cdot e^0_{23x} + r_y \cdot e^0_{23y} + r_z \cdot e^0_{23z} + r^0_x \cdot e_{23x} + r^0_y \cdot e_{23y} + r^0_z \cdot e_{23z} = 0$$
$$r_x \cdot e^0_{23x} + r_y \cdot 0 + r_z \cdot 0 + r^0_x \cdot 0 + r^0_y \cdot 0 + r^0_z \cdot 1 = 0$$
$$r_x \cdot e^0_{24x} + r_y \cdot e^0_{24y} + r_z \cdot e^0_{24z} + r^0_x \cdot e_{24x} + r^0_y \cdot e_{24y} + r^0_z \cdot e_{24z} = 0$$
$$r_x \cdot 0 + r_y \cdot 0 + r_z \cdot (-1) + r^0_x \cdot 0 + r^0_y \cdot 0 + r^0_z \cdot 0 = 0$$
$$r_x \cdot e^0_{25x} + r_y \cdot e^0_{25y} + r_z \cdot e^0_{25z} + r^0_x \cdot e_{25x} + r^0_y \cdot e_{25y} + r^0_z \cdot e_{25z} = 0$$
$$r_x \cdot e^0_{25x} + r_y \cdot 0 + r_z \cdot 0 + r^0_x \cdot 0 + r^0_y \cdot 0 + r^0_z \cdot 1 = 0$$

We can obtain wrench with $\boldsymbol{R}_2$ (0, 0, 0, 1, 0, 0) coordinates from the equation system.

Consider the third kinematic chain. We have four unit vectors. Reciprocation conditions for the Plucker coordinates of the unit screws and wrench can be written as four equations. For the solution to be found it is necessary to get two unknowns and consequently two wrenches:

$$r_x \cdot e^0_{32x} + r_y \cdot e^0_{32y} + r_z \cdot e^0_{32z} + r^0_x \cdot e_{32x} + r^0_y \cdot e_{32y} + r^0_z \cdot e_{32z} = 0$$
$$r_x \cdot 0 + r_y \cdot 0 + r_z \cdot 1 + r^0_x \cdot 0 + r^0_y \cdot 0 + r^0_z \cdot 0 = 0$$
$$r_x \cdot e^0_{33x} + r_y \cdot e^0_{33y} + r_z \cdot e^0_{33z} + r^0_x \cdot e_{33x} + r^0_y \cdot e_{33y} + r^0_z \cdot e_{33z} = 0$$
$$r_x \cdot 0 + r_y \cdot e^0_{33y} + r_z \cdot 0 + r^0_x \cdot (-1) + r^0_y \cdot 0 + r^0_z \cdot 0 = 0$$
$$r_x \cdot e^0_{34x} + r_y \cdot e^0_{34y} + r_z \cdot e^0_{34z} + r^0_x \cdot e_{34x} + r^0_y \cdot e_{34y} + r^0_z \cdot e_{34z} = 0$$
$$r_x \cdot 1 + r_y \cdot 0 + r_z \cdot 0 + r^0_x \cdot 0 + r^0_y \cdot 0 + r^0_z \cdot 0 = 0$$
$$r_x \cdot e^0_{35x} + r_y \cdot e^0_{35y} + r_z \cdot e^0_{35z} + r^0_x \cdot e_{35x} + r^0_y \cdot e_{35y} + r^0_z \cdot e_{35z} = 0$$
$$r_x \cdot 0 + r_y \cdot e^0_{35y} + r_z \cdot 0 + r^0_x \cdot (-1) + r^0_y \cdot 0 + r^0_z \cdot 0 = 0$$

Two wrenches with $\boldsymbol{R}_3$ (0, 0, 0, 0, 1, 0); $\boldsymbol{R}_4$ (0, 0, 0, 0, 0, 1) coordinates are obtained from the equation system.

Wrench coordinates can be presented as a corresponding $\boldsymbol{R}$ (6 × 4) matrix

$$\boldsymbol{R} = \begin{pmatrix} 0 & 0 & 0 & 0 & 0 & 1 \\ 0 & 0 & 0 & 1 & 0 & 0 \\ 0 & 0 & 0 & 0 & 1 & 0 \\ 0 & 0 & 0 & 0 & 0 & 1 \end{pmatrix}$$

This matrix shows that there are only three independent wrenches. These are the screws of infinity pitch (moments) of the infinity great characteristic quantity (moments) constraining the rotation.

Consider the first kinematic chain apart. Translate it about reference axis and get the Plucker coordinates of the chain (Fig. 6.9).

This kinematic chain duplicates the kinematic chain of the spatial 4-DOF mechanism exactly. Hence, we have the matrix with coordinates $\boldsymbol{E}_{11}$ $(1,0,0,0,0,0)$, $\boldsymbol{E}_{12}$ $(0,0,0,1,0,0)$, $\boldsymbol{E}_{13}$ $(0,-1,0,0,0,e_{13z}^{0})$, $\boldsymbol{E}_{14}$ $(0,0,0,e_{14x}^{0}, e_{14y}^{0},0)$, $\boldsymbol{E}_{15}$ $(0,-1,0,0,0,e_{15z}^{0})$, $\boldsymbol{E}_{16}$ $(1,0,0,0,0,e_{16z}^{0})$.

We can obtain the matrix of unit screws coordinates

$$\boldsymbol{E} = \begin{pmatrix} 1 & 0 & 0 & 0 & 0 & 0 \\ 0 & 0 & 0 & 1 & 0 & 0 \\ 0 & -1 & 0 & 0 & 0 & e_{13z}^{0} \\ 0 & 0 & 0 & e_{14x}^{0} & e_{15y}^{0} & 0 \\ 0 & -1 & 0 & 0 & 0 & e_{15z}^{0} \\ 1 & 0 & 0 & 0 & 0 & e_{16z}^{0} \end{pmatrix}$$

Repeat the all algorithm given above. As a result, we obtain the following coefficients:

$$a = 1, b = -c, c = \frac{e_{16z}^{0}}{e_{15}^{0} - e_{13z}^{0}}$$

Therefore, we can find stated coefficients that evaluate the speeds in the corresponding kinematic pairs. There is the internal mobility in the chain. Therefore, it is possible to translate rotation from the rotary actuator to the rotation kinematic pair coupled with the output link.

There are internal mobilities in the two kinematic chains—the rotation is translated from the rotary actuated to the final rotation kinematic pair. There are two

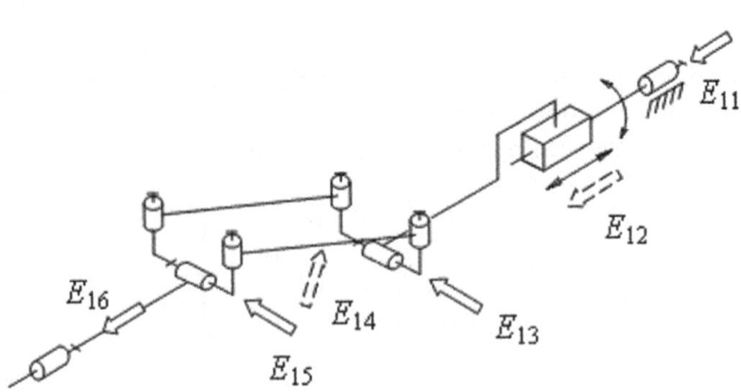

Fig. 6.9 First kinematic chain of the 5-DOF mechanism

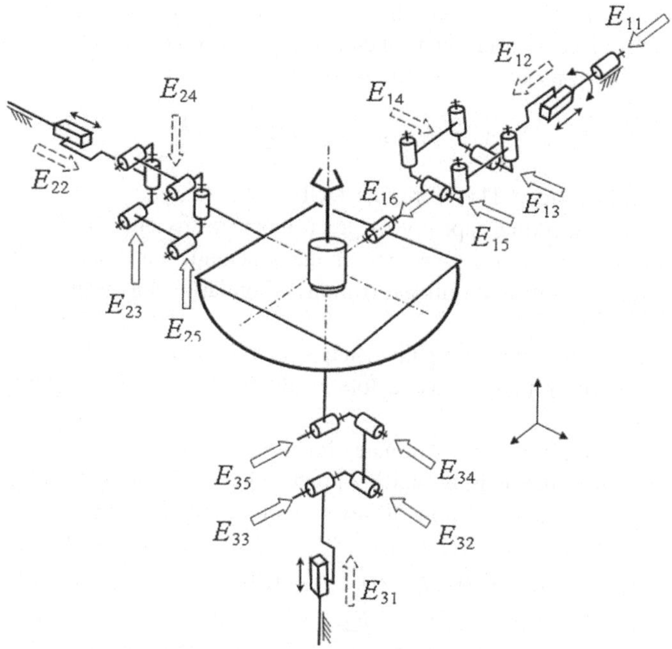

Fig. 6.10 Schematic structure of the 5-DOF parallel mechanism

DOF necessary for the partial mechanism which is arranged between output and final links. Let us make sure of it.

In according with the Dobrovolysky formula we have

$$W = 3 \cdot (n - 1) - 2 \cdot p_5 = 3 \cdot (5 - 1) - 2 \cdot 5 = 2$$

Thus, the partial spherical mechanism has two DOF. So the output link moving in three coordinates has three translational DOF. And the final link has two rotational DOF with respect to the output link.

We can make such analysis for modified mechanism for which the kinematic pair unit screws are presented (Fig. 6.10).

6.2.3 *Schematic Structure of the 6-DOF Parallel Mechanism*

Consider the 6-DOF mechanism after having added three rotational motion drivers in three kinematic pairs to the original scheme of the ORTHOGLIDE mechanism (Fig. 6.6) [4].

In this case examine the same links that before, excluding intermediate parallelogram links. Twelve rotational kinematic pairs of three parallelograms are substituted for three prismatic kinematic pairs. We get

$$W = 6 \cdot (17 - 1) - 5 \cdot 18 = 6$$

This result is correct. Therefore, the DOF equals 6.

To determine working capacity of given mechanisms one should consider the Plucker coordinates of basis vectors of the kinematic pairs axes. This method permits to identify potential critical positions (singularities) as well as uncontrolled mobility.

In this case each kinematic pair matches its axis basis vector and the Pluecker coordinates matrix of these basis vectors is compiled. The corresponding design is on the Fig. 6.11.

Thus, regard the Plucker coordinates for the unit screws of the kinematic pairs axes of mechanism that is in its starting point (Fig. 6.12). The kinematic pairs axes position is determined by units screws with coordinates: $\boldsymbol{E}_{11}\,(1,0,0,0,0,0)$, $\boldsymbol{E}_{12}\,(0,0,0,1,0,0)$, $\boldsymbol{E}_{13}\,(0,1,0,0,0,e^0_{13z})$, $\boldsymbol{E}_{14}\,(0,0,0,0,-1,0)$, $\boldsymbol{E}_{15}\,(0,1,0,0,0,e^0_{15z})$, $\boldsymbol{E}_{16}\,(1,0,0,0,0,0)$, $\boldsymbol{E}_{21}\,(0,1,0,0,0,0)$, $\boldsymbol{E}_{22}\,(0,0,0,0,1,0)$, $\boldsymbol{E}_{23}\,(0,0,-1,e^0_{23x},0,0)$, $\boldsymbol{E}_{24}\,(0,0,0,0,0,1)$, $\boldsymbol{E}_{25}\,(0,0,-1,e^0_{25x},0,0)$, $\boldsymbol{E}_{26}\,(0,1,0,0,0,0)$, $\boldsymbol{E}_{31}\,(0,0,1,0,0,0)$, $\boldsymbol{E}_{32}\,(0,0,0,0,0,1)$, $\boldsymbol{E}_{33}\,(-1,0,0,0,e^0_{33y},0)$, $\boldsymbol{E}_{34}\,(0,0,0,1,0,0)$, $\boldsymbol{E}_{35}\,(-1,0,0,0,e^0_{35y},0)$, $\boldsymbol{E}_{36}\,(0,0,1,0,0,0)$.

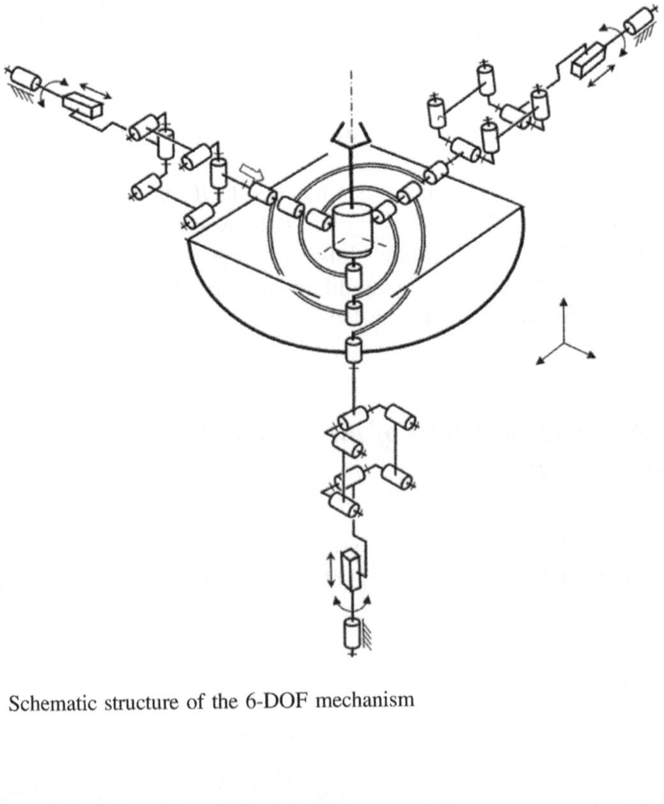

Fig. 6.11 Schematic structure of the 6-DOF mechanism

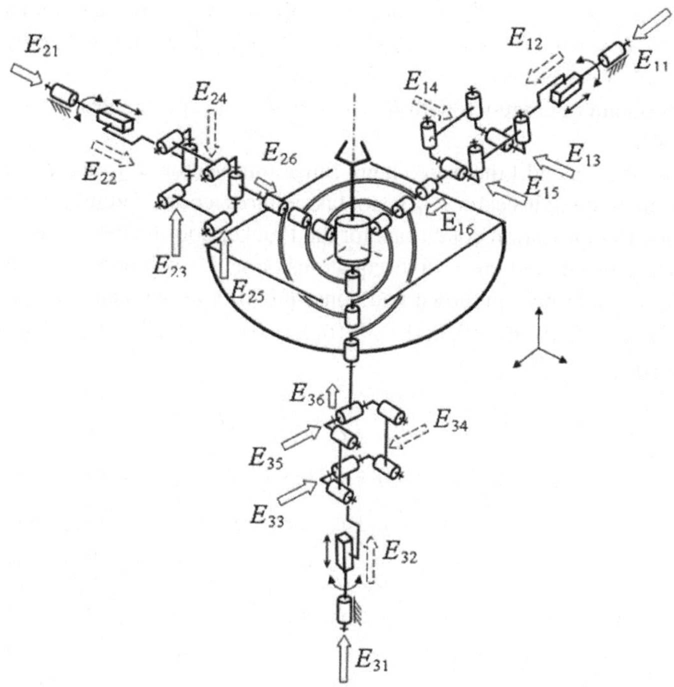

Fig. 6.12 Schematic structure of the 6-DOF parallel mechanism

Examine the first kinematic chain. Since unit vectors $\boldsymbol{E}_{16}$ and $\boldsymbol{E}_{16}$ are on the same axis then the unit vector $\boldsymbol{E}_{16}$ is not taken into account. Therefore we have five unit vectors. Reciprocation conditions for the Plucker coordinates of the unit and wrench can be written as five equations. For the solution to be found it is necessary to obtain one unknown and consequently one wrench.

$$r_x \cdot e^0_{11x} + r_y \cdot e^0_{11y} + r_z \cdot e^0_{11z} + r^0_x \cdot e_{11x} + r^0_y \cdot e_{11y} + r^0_z \cdot e_{11z} = 0$$
$$r_x \cdot 0 + r_y \cdot 0 + r_z \cdot 0 + r^0_x \cdot 1 + r^0_y \cdot 0 + r^0_z \cdot 0 = 0$$
$$r_x \cdot e^0_{12x} + r_y \cdot e^0_{12y} + r_z \cdot e^0_{12z} + r^0_x \cdot e_{12x} + r^0_y \cdot e_{12y} + r^0_z \cdot e_{12z} = 0$$
$$r_x \cdot 1 + r_y \cdot 0 + r_z \cdot 0 + r^0_x \cdot 0 + r^0_y \cdot 0 + r^0_z \cdot 0 = 0$$
$$r_x \cdot e^0_{13x} + r_y \cdot e^0_{13y} + r_z \cdot e^0_{13z} + r^0_x \cdot e_{13x} + r^0_y \cdot e_{13y} + r^0_z \cdot e_{13z} = 0$$
$$r_x \cdot 0 + r_y \cdot 0 + r_z \cdot e^0_{13z} + r^0_x \cdot 0 + r^0_y \cdot 1 + r^0_z \cdot 0 = 0$$
$$r_x \cdot e^0_{14x} + r_y \cdot e^0_{14y} + r_z \cdot e^0_{14z} + r^0_x \cdot e_{14x} + r^0_y \cdot e_{14y} + r^0_z \cdot e_{14z} = 0$$
$$r_x \cdot 0 + r_y \cdot (-1) + r_z \cdot 0 + r^0_x \cdot 0 + r^0_y \cdot 0 + r^0_z \cdot 0 = 0$$
$$r_x \cdot e^0_{15x} + r_y \cdot e^0_{15y} + r_z \cdot e^0_{15z} + r^0_x \cdot e_{15x} + r^0_y \cdot e_{15y} + r^0_z \cdot e_{15z} = 0$$
$$r_x \cdot 0 + r_y \cdot 0 + r_z \cdot e^0_{15z} + r^0_x \cdot 0 + r^0_y \cdot 1 + r^0_z \cdot 0 = 0$$

where e_{ijx}, e_{ijy}, e_{ijz}, e^0_{ijx}, e^0_{ijy}, e^0_{ijz} are the unit screw vector and moment parts, respectively. (i = 1–3; j = 1–3); r_{ijx}, r_{ijy}, r_{ijz}, r^0_{ijx}, r^0_{ijy}, r^0_{ijz} are the wrench vector and moment parts, respectively (i = 1–3; j = 1–3).

We can obtain one wrench with $\boldsymbol{R}_1$ (0, 0, 0, 0, 0, 1) coordinates from the system of equations.

Consider the second kinematic chain. Since unit vectors $\boldsymbol{E}_{21}$ and $\boldsymbol{E}_{26}$ are on the same axis, then the unit vector $\boldsymbol{E}_{26}$ is not taken into account. Therefore we have five unit vectors. Reciprocation conditions for the Plucker coordinates of the unit screws and wrench can be written as five equations. For the solution to be found it is necessary to obtain one unknown and consequently one wrench.

$\boldsymbol{E}_{21}$ (0, 1, 0, 0, 0, 0), $\boldsymbol{E}_{22}$ (0, 0, 0, 0, 1, 0), $\boldsymbol{E}_{23}$ $(0, 0, -1, e^0_{23x}, 0, 0)$, $\boldsymbol{E}_{24}$ (0, 0, 0, 0, 0, 1), $\boldsymbol{E}_{25}$ $(0, 0, -1, e^0_{25x}, 0, 0)$

$$
\begin{aligned}
r_x \cdot e^0_{21x} + r_y \cdot e^0_{21y} + r_z \cdot e^0_{21z} + r^0_x \cdot e_{21x} + r^0_y \cdot e_{21y} + r^0_z \cdot e_{21z} = 0 \\
r_x \cdot 0 + r_y \cdot 0 + r_z \cdot 0 + r^0_x \cdot 0 + r^0_y \cdot 1 + r^0_z \cdot 0 = 0 \\
r_x \cdot e^0_{22x} + r_y \cdot e^0_{22y} + r_z \cdot e^0_{22z} + r^0_x \cdot e_{22x} + r^0_y \cdot e_{22y} + r^0_z \cdot e_{22z} = 0 \\
r_x \cdot 0 + r_y \cdot 1 + r_z \cdot 0 + r^0_x \cdot 0 + r^0_y \cdot 0 + r^0_z \cdot 0 = 0 \\
r_x \cdot e^0_{23x} + r_y \cdot e^0_{23y} + r_z \cdot e^0_{23z} + r^0_x \cdot e_{23x} + r^0_y \cdot e_{23y} + r^0_z \cdot e_{23z} = 0 \\
r_x \cdot e^0_{23x} + r_y \cdot 0 + r_z \cdot 0 + r^0_x \cdot 0 + r^0_y \cdot 0 + r^0_z \cdot (-1) = 0 \\
r_x \cdot e^0_{24x} + r_y \cdot e^0_{24y} + r_z \cdot e^0_{24z} + r^0_x \cdot e_{24x} + r^0_y \cdot e_{24y} + r^0_z \cdot e_{24z} = 0 \\
r_x \cdot 0 + r_y \cdot 0 + r_z \cdot 1 + r^0_x \cdot 0 + r^0_y \cdot 0 + r^0_z \cdot 0 = 0 \\
r_x \cdot e^0_{25x} + r_y \cdot e^0_{25y} + r_z \cdot e^0_{25z} + r^0_x \cdot e_{25x} + r^0_y \cdot e_{25y} + r^0_z \cdot e_{25z} = 0 \\
r_x \cdot e^0_{25x} + r_y \cdot 0 + r_z \cdot 0 + r^0_x \cdot 0 + r^0_y \cdot 0 + r^0_z \cdot (-1) = 0
\end{aligned}
$$

We can obtain wrench with $\boldsymbol{R}_2$ (0, 0, 0, 1, 0, 0) coordinates from the equation system.

Consider the third kinematic chain. Since unit vectors $\boldsymbol{E}_{31}$ and $\boldsymbol{E}_{36}$ are on the same axis, then the unit vector $\boldsymbol{E}_{36}$ is not taken into account. Therefore, we have five unit vectors. Reciprocation conditions for the Plucker coordinates of the unit screws and wrench can be written as five equations. For the solution to be found it is necessary to obtain one unknown and consequently one wrench

$$r_x \cdot e^0_{31x} + r_y \cdot e^0_{31y} + r_z \cdot e^0_{31z} + r^0_x \cdot e_{31x} + r^0_y \cdot e_{31y} + r^0_z \cdot e_{31z} = 0$$
$$r_x \cdot 0 + r_y \cdot 0 + r_z \cdot 0 + r^0_x \cdot 0 + r^0_y \cdot 0 + r^0_z \cdot 1 = 0$$
$$r_x \cdot e^0_{32x} + r_y \cdot e^0_{32y} + r_z \cdot e^0_{32z} + r^0_x \cdot e_{32x} + r^0_y \cdot e_{32y} + r^0_z \cdot e_{32z} = 0$$
$$r_x \cdot 0 + r_y \cdot 0 + r_z \cdot 1 + r^0_x \cdot 0 + r^0_y \cdot 0 + r^0_z \cdot 0 = 0$$
$$r_x \cdot e^0_{33x} + r_y \cdot e^0_{33y} + r_z \cdot e^0_{33z} + r^0_x \cdot e_{33x} + r^0_y \cdot e_{33y} + r^0_z \cdot e_{33z} = 0$$
$$r_x \cdot 0 + r_y \cdot e^0_{33y} + r_z \cdot 0 + r^0_x \cdot (-1) + r^0_y \cdot 0 + r^0_z \cdot 0 = 0$$
$$r_x \cdot e^0_{34x} + r_y \cdot e^0_{34y} + r_z \cdot e^0_{34z} + r^0_x \cdot e_{34x} + r^0_y \cdot e_{34y} + r^0_z \cdot e_{34z} = 0$$
$$r_x \cdot 1 + r_y \cdot 0 + r_z \cdot 0 + r^0_x \cdot 0 + r^0_y \cdot 0 + r^0_z \cdot 0 = 0$$
$$r_x \cdot e^0_{35x} + r_y \cdot e^0_{35y} + r_z \cdot e^0_{35z} + r^0_x \cdot e_{35x} + r^0_y \cdot e_{35y} + r^0_z \cdot e_{35z} = 0$$
$$r_x \cdot 0 + r_y \cdot e^0_{35y} + r_z \cdot 0 + r^0_x \cdot (-1) + r^0_y \cdot 0 + r^0_z \cdot 0 = 0$$

We can obtain wrench with $\boldsymbol{R}_3$ (0, 0, 0, 0, 1, 0) coordinates from the equation system.

Wrench coordinates can be represent as an corresponding matrix $\boldsymbol{R}$ (6 × 3):

$$\boldsymbol{R} = \begin{pmatrix} 0 & 0 & 0 & 0 & 0 & 1 \\ 0 & 0 & 0 & 1 & 0 & 0 \\ 0 & 0 & 0 & 0 & 1 & 0 \end{pmatrix}$$

This matrix shows that there are only three independent wrench. These are the screws of infinity pitch (moments) of the infinity great characteristic quantity (moments) constraining the rotation.

Consider the first kinematic chain apart. Translate it about reference axis and get the Plucker coordinates of the chain.

This kinematic chain duplicates the kinematic chain of the spatial 4-DOF and 5-DOF mechanism exactly. Hence, it has coordinates: $\boldsymbol{E}_{11}$ $(1,0,0,0,0,0)$, $\boldsymbol{E}_{12}$ $(0,0,0,1,0,0)$, $\boldsymbol{E}_{13}$ $(0,-1,0,0,0,e^0_{13z})$, $\boldsymbol{E}_{14}$ $(0,0,0,e^0_{14x},e^0_{14y},0)$, $\boldsymbol{E}_{15}$ $(0,-1,0,0,0,e^0_{15z})$, $\boldsymbol{E}_{16}$ $(1,0,0,0,0,e^0_{16z})$ (Fig. 6.13).

We get the matrix of unit screws coordinates:

$$\boldsymbol{E} = \begin{pmatrix} 1 & 0 & 0 & 0 & 0 & 0 \\ 0 & 0 & 0 & 1 & 0 & 0 \\ 0 & -1 & 0 & 0 & 0 & e^0_{13z} \\ 0 & 0 & 0 & e^0_{14x} & e^0_{15y} & 0 \\ 0 & -1 & 0 & 0 & 0 & e^0_{15z} \\ 1 & 0 & 0 & 0 & 0 & e^0_{16z} \end{pmatrix}$$

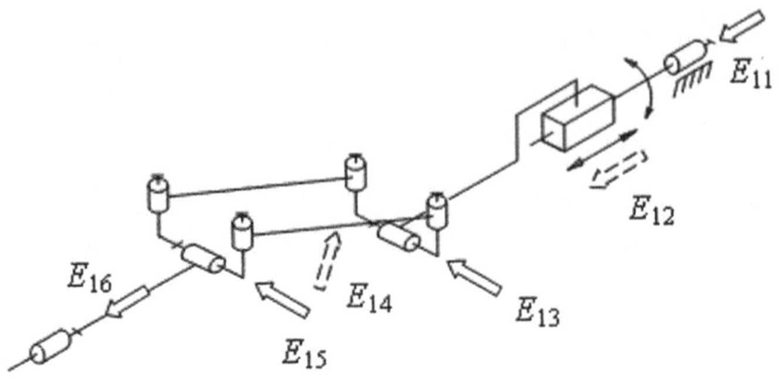

Fig. 6.13 First kinematic chain of the 4-DOF mechanism

Repeat the all algorithm given above. As a result, we obtain the following coefficients

$$a = 1, b = -c, c = \frac{e^0_{16z}}{e^0_{15} - e^0_{13z}},$$

Therefore we can find stated coefficients that evaluate the speeds in the corresponding kinematic pairs. There is the internal mobility in the chain. Therefore it is possible to translate the rotation from rotary actuator to the rotation kinematic pair coupled with the output link.

There are internal mobilities in the three kinematic chains—the rotation is translated from the rotary actuator to the final rotation kinematic pair. There are three DOF required for the partial mechanism, which is arranged between the output and final links. Let us make sure of it.

In according with the Dobrovolysky formula we have

$$W = 3 \cdot (n - 1) - 2 \cdot p_5 = 3 \cdot (6 - 1) - 2 \cdot 6 = 3$$

Thus, the partial spherical mechanism has three DOF. The output link moving in three coordinates has three translational DOF. The final link has two rotational DOF with respect to the output link.

New kinematic structures of the 4-DOF, the 5-DOF and 6-DOF spatial mechanisms are proposed as a result of the type synthesis of development mechanisms using the Malyshev and the Chebyshev classical formulas for the spatial and plain mechanisms.

The working capacity of the development mechanisms was determined by means of the screw method.

6.3 Solving the Kinematic Problem for Parallel Structure Mechanisms (PMs) with Translational and Rotational Motion Decomposition

Here the kinematic problem being of high priority for analysis, synthesis and control is regarded.

6.3.1 *Solving the Kinematic Problem for the Translational Mechanism*

Solving the inverse kinematics problem for PM with output link translational motion.

Let us consider the inverse kinematics solution (on positions) for PM with linear motors arranging Cartesian system direction. This structure corresponds to the mechanisms studied in the previous chapter, where translational motions have already been determined.

Since each chain has the revolute joint parallel mechanism, two rotation pairs, and a linear motor, the translational motions are assumed to be caused. This structure can be substituted with a "pyramid" mechanism (Fig. 6.14) that can be applied for ORTHOGLIDE's type structures (Fig. 6.15). Substitute mechanism comes from the initial one so that all kinematics chains are mentally shifted to the output link centre.

The intervals characterizing the B_1, B_2, B_3 point positions are generalized coordinates. The A point coordinates are the absolute coordinates of the output link.

The aim of the inverse kinematic problem solution is to search the input links position values (x_{B1}, y_{B2}, z_{B3}), with tool set point coordinates (x_A; y_A; z_A) of the manipulator being known. In general, the inverse kinematic problem can have several solutions corresponding with one tool position.

The point A is in the xyz coordinates and it is not precisely the point of reference O. The task is to determine the coordinates x_{B1}, y_{B2}, z_{B3}, i.e. what distance L_x^*, L_y^*, L_x^* the input mechanism links are supposed to be replaced. The coordinates of the point A are (x_A; y_A; z_A). To learn the distance the mechanism input links have been displaced relating to the point of reference the following equations can be generated:

$$
\begin{aligned}
(x_A - x_{B1})^2 + (y_A - y_{B1})^2 + (z_A - z_{B1})^2 - L^2 &= 0\\
(x_A - x_{B2})^2 + (y_A - y_{B2})^2 + (z_A - z_{B2})^2 - L^2 &= 0\\
(x_A - x_{B3})^2 + (y_A - y_{B3})^2 + (z_A - z_{B3})^2 - L^2 &= 0
\end{aligned}
$$

Here L is the intermediate link length of each chain. Appoint coordinates x_{B1}, y_{B2}, z_{B3}, as q_1, q_2, q_3. Appoint coordinates x_{B1}, y_{B2}, z_{B3}, as q_1, q_2, q_3. Since the

Fig. 6.14 "Pyramid" mechanism is the 3-DOF mechanism computational structure

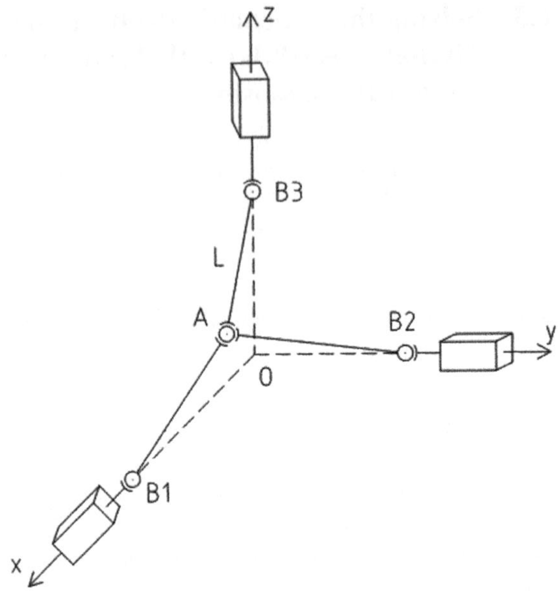

Fig. 6.15 3-DOF translational-guiding mechanism

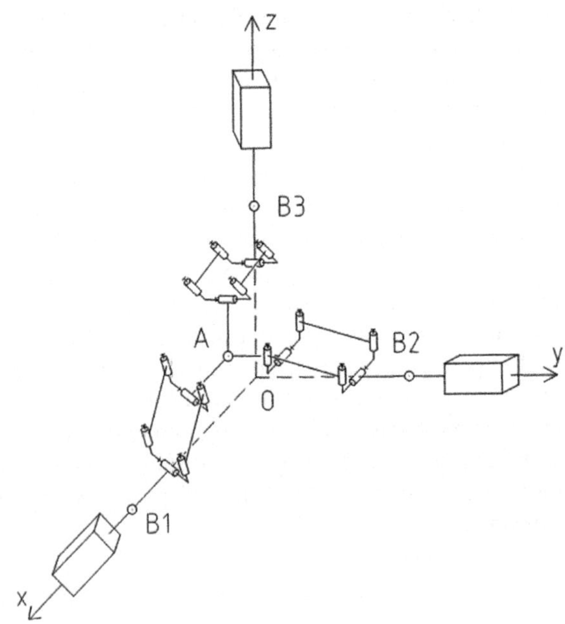

corresponding input links are along the coordinate axes it follows that y_{B1}^2, z_{B1}^2, x_{B2}^2, z_{B2}^2, x_{B3}^2, y_{B3}^2 coordinates become zero and we have the following equations:

$$(x_A - q_1)^2 + y_A^2 + z_A^2 - L^2 = 0$$
$$x_A^2 + (y_A - q_2)^2 + z_A^2 - L^2 = 0$$
$$x_A^2 + y_A^2 + (z_A - q_3)^2 - L^2 = 0$$

Three unknowns, i.e. q_1, q_2, q_3 should be found from three obtained independent equations.

The equations in two unknowns could be transformed into the quadratic equation form

$$A \cdot q_1^2 + B \cdot q_1 + C = 0$$

Then

$$q_1 = \frac{-B \pm \sqrt{B^2 - 4 \cdot A \cdot C}}{2 \cdot A},$$
$$B = -2 \cdot x_A,$$
$$C = x_A^2 + y_A^2 + z_A^2 - L^2,$$
$$A \cdot q_2^2 + B \cdot q_2 + C = 0$$

In a similar way to determine q_2, q_3

$$q_2 = \frac{-B \pm \sqrt{B^2 - 4 \cdot A \cdot C}}{2 \cdot A},$$
$$B = -2 \cdot y_A,$$
$$C = x_A^2 + y_A^2 + z_A^2 - L^2$$

The direct kinematic problem solution for PM with translational motion of output link.

In the case of direct kinematic problem the input links location values (q_1, q_2, q_3) are considered to be known, with the tool set point position (x_A, y_A, z_A) being necessary to find. This problem can have several solutions corresponding different tool positions.

To find the point A coordinates (x_A; y_A; z_A). The y_{B1}^2, z_{B1}^2, x_{B2}^2, z_{B2}^2, x_{B3}^2, y_{B3}^2 coordinates become zero. Then x_A, y_A, z_A should be equal zero

$$x_A = \frac{-B \pm \sqrt{B^2 - 4 \cdot A \cdot C}}{2 \cdot A}$$

$$A = b^2 + d^2,$$

$$B = -2 \cdot q_1 + 2 \cdot a \cdot b + 2 \cdot c \cdot d,$$

$$C = q_1^2 + a^2 + c^2 - L^2$$

$$a = \frac{q_2^2 - q_1^2}{2 \cdot q_2},$$

$$b = \frac{q_1}{q_2}$$

$$y_A = a + b \cdot x_A,$$

$$z_A = c + d \cdot x_A$$

$$c = \frac{q_3^2 - q_1^2}{2 \cdot q_3},$$

$$d = \frac{q_1}{q_3}$$

6.3.2 *Translational Mechanism Velocity Problem Solution*

The velocity problem solution is represented by the equations of constraints derivation method.

The constraint equation for translational-guiding manipulator with three kinematic chains can be offered by the system of equations.

$$\begin{aligned} F_1 &= (x_A - q_1)^2 + y_A^2 + z_A^2 - L^2 = 0 \\ F_2 &= x_A^2 + (y_A - q_2)^2 + z_A^2 - L^2 = 0 \\ F_3 &= x_A^2 + y_A^2 + (z_A - q_3)^2 - L^2 = 0 \end{aligned} \tag{6.1}$$

The constraint equation for translational-guiding manipulator with three kinematic chains can be offered by the system of Eq. (6.1).

Thus we get the constraint equations between the input and output links angles of rotation.

Let us use analytical method for velocity problem to be solved. It is based on the Jacobian matrix study given in general terms:

$$A \cdot V = (-B) \cdot q_i \tag{6.2}$$

where A is the matrix containing partial derivatives of the implicit function with respect to x_A, y_A, z_A; B is the matrix containing partial derivatives of implicit function with respect to generalized coordinates; V is the output link velocity vector; q_i is the generalized velocites in revolute joints (input velocities).

Differentiate the constraint Eq. (6.1) of translational-guiding manipulator with three kinematic chains and obtain the following expression

$$
\begin{aligned}
&\tfrac{\partial F_1}{\partial x_A} = 2 \cdot x_A - 2 \cdot q_1; \quad \tfrac{\partial F_1}{\partial y_A} = 2 \cdot y_A; \quad \tfrac{\partial F_1}{\partial z_A} = 2 \cdot z_A; \quad \tfrac{\partial F_1}{\partial q_1} = 2 \cdot q_1 - 2 \cdot x_A \\
&\tfrac{\partial F_2}{\partial x_A} = 2 \cdot x_A; \quad \tfrac{\partial F_2}{\partial y_A} = 2 \cdot y_A - 2 \cdot q_2; \quad \tfrac{\partial F_2}{\partial z_A} = 2 \cdot z_A; \quad \tfrac{\partial F_2}{\partial q_2} = 2 \cdot q_2 - 2 \cdot y_A \\
&\tfrac{\partial F_3}{\partial x_A} = 2 \cdot z_A; \quad \tfrac{\partial F_3}{\partial y_A} = 2 \cdot y_A; \quad \tfrac{\partial F_3}{\partial z_A} = 2 \cdot z_A - 2 \cdot q_3; \quad \tfrac{\partial F_3}{\partial q_3} = 2 \cdot q_3 - 2 \cdot z_A
\end{aligned}
$$

In general terms the (6.2) equation is given by:

$$
\begin{pmatrix} 2 \cdot x_A - 2 \cdot q_1 & 2 \cdot y_A & 2 \cdot z_A \\ 2 \cdot y_A & 2 \cdot y_A - 2 \cdot q_2 & 2 \cdot z_A \\ 2 \cdot z_A & 2 \cdot y_A & 2 \cdot z_A - 2 \cdot q_3 \end{pmatrix} \cdot \begin{pmatrix} V_x \\ V_y \\ V_z \end{pmatrix} =
- \begin{pmatrix} 2 \cdot q_1 - 2 \cdot x_A & 0 & 0 \\ 0 & 2 \cdot q_2 - 2 \cdot y_A & 0 \\ 0 & 0 & 2 \cdot q_3 - 2 \cdot z_A \end{pmatrix} \cdot \begin{pmatrix} q_1 \\ q_2 \\ q_3 \end{pmatrix}
$$

6.3.3 Spherical Mechanism Kinematic Problem

Consider the spherical part of the 6-DOF synthesized mechanism (Fig. 6.16) consisting of three kinematic chains whose axes at right angle. Thus each output link is coupled with rotational motion motor. The output link consists of two intersecting spheres and a tool, which rotates in the x, y, z axes meeting at the O point.

The output coordinates are the α, β, γ platform rotation angles about x, y, z axes respectively. The generalized coordinates are the φ_{11}, φ_{21}, φ_{31} being the rotation angles of the first, second, and third input links respectively.

To determine velocities and critical positions of the manipulator, the kinematic problem is bound to be solved, i.e. the relation between the input and output coordinates is necessary to be found. Let us use analytical method to perform the computation.

The output link is turning consequently about the x, y, z axes. Matrixes specifying these turns in the transition from moving coordinate system to fixed coordinate system are in the form

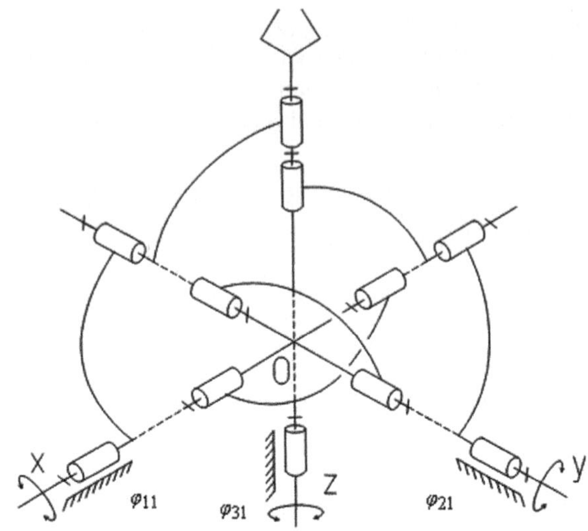

Fig. 6.16 Spherical part of the 6-DOF mechanism

$$A_1 = \begin{pmatrix} 1 & 0 & 0 \\ 0 & \cos\alpha & -\sin\alpha \\ 0 & \sin\alpha & \cos\alpha \end{pmatrix}, A_2 = \begin{pmatrix} \cos\beta & 0 & \sin\beta \\ 0 & 1 & 0 \\ -\sin\beta & 0 & \cos\beta \end{pmatrix},$$

$$A_3 = \begin{pmatrix} \cos\gamma & -\sin\gamma & 0 \\ \sin\gamma & \cos\gamma & 0 \\ 0 & 0 & 1 \end{pmatrix}$$

Matrix, showing the output link transition from moving coordinate system to fixed coordinate system is in the form

$$A = A_3 \cdot A_2 \cdot A_1$$

$$A = \begin{pmatrix} \cos\beta\cdot\cos\gamma & \cos\gamma\cdot\sin\alpha\cdot\sin\beta - \cos\alpha\cdot\sin\gamma & \sin\alpha\cdot\sin\gamma + \cos\alpha\cdot\cos\gamma\cdot\sin\beta \\ \cos\beta\cdot\sin\gamma & \cos\alpha\cdot\cos\gamma + \sin\alpha\cdot\sin\beta\cdot\sin\gamma & \cos\alpha\cdot\sin\beta\cdot\sin\gamma - \cos\gamma\cdot\sin\alpha \\ -\sin\beta & \cos\beta\cdot\sin\alpha & \cos\alpha\cdot\cos\beta \end{pmatrix}$$

In the first kinematic chain the input link turns about the axes in the following order x, y, z. Corresponding rotation matrixes are given by

$$B'_1 = \begin{pmatrix} 1 & 0 & 0 \\ 0 & \cos\varphi_{11} & -\sin\varphi_{11} \\ 0 & \sin\varphi_{11} & \cos\varphi_{11} \end{pmatrix}, B'_2 = \begin{pmatrix} \cos\varphi_{21} & 0 & \sin\varphi_{21} \\ 0 & 1 & 0 \\ -\sin\varphi_{21} & 0 & \cos\varphi_{21} \end{pmatrix},$$

$$B'_3 = \begin{pmatrix} \cos\varphi_{31} & -\sin\varphi_{31} & 0 \\ \sin\varphi_{31} & \cos\varphi_{31} & 0 \\ 0 & 0 & 1 \end{pmatrix}$$

where φ_{11} is the rotation angle of the input link; φ_{21} is the rotation angle of the second link; φ_{31} is the rotation angle of the third link.

Matrix of the input link transition from moving coordinate system to fixed coordinate system may be offered as

$$B' = B'_1 \cdot B'_2 \cdot B'_3$$

$$B' = \begin{pmatrix} \cos\varphi_{12}\cdot\cos\varphi_{13} & -\cos\varphi_{12}\cdot\sin\varphi_{13} & \sin\varphi_{12} \\ \cos\varphi_{11}\cdot\sin\varphi_{13}+\cos\varphi_{13}\cdot\sin\varphi_{11}\cdot\sin\varphi_{12} & \cos\varphi_{11}\cdot\cos\varphi_{13}-\sin\varphi_{11}\cdot\sin\varphi_{12}\cdot\sin\varphi_{13} & -\cos\varphi_{12}\cdot\sin\varphi_{11} \\ \sin\varphi_{11}\cdot\sin\varphi_{13}-\cos\varphi_{11}\cdot\cos\varphi_{13}\cdot\sin\varphi_{12} & \cos\varphi_{13}\cdot\sin\varphi_{11}+\cos\varphi_{11}\cdot\sin\varphi_{12}\cdot\sin\varphi_{13} & \cos\varphi_{11}\cdot\cos\varphi_{12} \end{pmatrix}$$

The kinematic pair vector of the first chain, coupled with the output link, has the following coordinates $\begin{pmatrix} 0 \\ 0 \\ 1 \end{pmatrix}$.

Substitute the obtained A and B matrix values in the constraint equation

$$A \cdot \begin{pmatrix} 0 \\ 0 \\ 1 \end{pmatrix} = B' \cdot \begin{pmatrix} 0 \\ 0 \\ 1 \end{pmatrix}$$

We get

$$\begin{pmatrix} \sin\beta \\ -\cos\beta\cdot\sin\alpha \\ \cos\alpha\cdot\cos\beta \end{pmatrix} = \begin{pmatrix} \sin\varphi_{12} \\ -\cos\varphi_{12}\cdot\sin\varphi_{11} \\ \cos\varphi_{11}\cdot\cos\varphi_{12} \end{pmatrix},$$

out of which the φ_{11} angle can be expressed through the α, β, and γ

$$\mathrm{tg}\phi_{11} = \frac{\cos\beta \cdot \sin\alpha}{\cos\alpha \cdot \sin\alpha}$$

Matrix of transition from moving coordinate system to fixed coordinate system for the second chain may be offered as

$$B' = B_2' \cdot B_1' \cdot B_3'$$

$$B'' = \begin{pmatrix} \cos\varphi_{21}\cdot\cos\varphi_{33}+\sin\varphi_{21}\cdot\sin\varphi_{22}\cdot\sin\varphi_{23} & \cos\varphi_{23}\cdot\sin\varphi_{21}\cdot\sin\varphi_{22}-\cos\varphi_{21}\cdot\sin\varphi_{23} & \cos\varphi_{22}\cdot\sin\varphi_{21} \\ \cos\varphi_{22}\cdot\sin\varphi_{23} & \cos\varphi_{22}\cdot\cos\varphi_{23} & -\sin\varphi_{22} \\ \cos\varphi_{21}\cdot\sin\varphi_{22}\cdot\sin\varphi_{23}-\cos\varphi_{23}\cdot\sin\varphi_{21} & \sin\varphi_{21}\cdot\sin\varphi_{23}+\cos\varphi_{21}\cdot\cos\varphi_{23}\cdot\sin\varphi_{22} & \cos\varphi_{21}\cdot\cos\varphi_{22} \end{pmatrix}$$

The second chain output link vector has the following coordinates $\begin{pmatrix} 0 \\ 0 \\ 1 \end{pmatrix}$.

Substitute the obtained A and B'' matrix values in the constraint equation

$$A \cdot \begin{pmatrix} 0 \\ 0 \\ 1 \end{pmatrix} = B'' \cdot \begin{pmatrix} 0 \\ 0 \\ 1 \end{pmatrix}$$

We get

$$\begin{pmatrix} \sin\beta \\ -\cos\beta\cdot\sin\alpha \\ \cos\alpha\cdot\cos\beta \end{pmatrix} = \begin{pmatrix} \cos\varphi_{22}\cdot\sin\varphi_{21} \\ -\sin\varphi_{22} \\ \cos\varphi_{21}\cdot\cos\varphi_{22} \end{pmatrix},$$

out of which the φ_{21} angle can be expressed through the α, β, and γ

$$\mathrm{tg}\varphi_{21} = \frac{\sin\beta}{\cos\alpha\cdot\cos\beta}$$

Matrix of the transition from moving coordinate system to fixed coordinate system for the third chain is offered as

$$B' = B_3' \cdot B_1' \cdot B_2'$$

$$B^{//} = \begin{pmatrix} \cos\varphi_{31}\cdot\cos\varphi_{33} - \sin\varphi_{31}\cdot\sin\varphi_{32}\cdot\sin\varphi_{33} & -\cos\varphi_{32}\cdot\sin\varphi_{31} & \cos\varphi_{31}\cdot\sin\varphi_{33} + \cos\varphi_{33}\cdot\sin\varphi_{31}\cdot\sin\varphi_{32} \\ \cos\varphi_{33}\cdot\sin\varphi_{31} + \cos\varphi_{31}\cdot\sin\varphi_{32}\cdot\sin\varphi_{33} & \cos\varphi_{31}\cdot\cos\varphi_{32} & \sin\varphi_{31}\cdot\sin\varphi_{33} - \cos\varphi_{31}\cdot\cos\varphi_{33}\cdot\sin\varphi_{32} \\ -\cos\varphi_{32}\cdot\sin\varphi_{33} & \sin\varphi_{32} & \cos\varphi_{32}\cdot\cos\varphi_{33} \end{pmatrix}$$

The third chain output link vector has the following coordinates $\begin{pmatrix} 0 \\ 0 \\ 1 \end{pmatrix}$.

Substitute the obtained A and B''' matrix values in the constraint equation

$$A \cdot \begin{pmatrix} 0 \\ 1 \\ 0 \end{pmatrix} = B''' \cdot \begin{pmatrix} 0 \\ 1 \\ 0 \end{pmatrix}$$

We get

$$\begin{pmatrix} -\cos\beta\cdot\sin\gamma \\ \cos\alpha\cdot\cos\gamma - \sin\alpha\cdot\sin\beta\cdot\sin\gamma \\ \cos\gamma\cdot\sin\alpha + \cos\alpha\cdot\sin\beta\cdot\sin\gamma \end{pmatrix} = \begin{pmatrix} -\cos\varphi_{32}\cdot\sin\varphi_{31} \\ \cos\varphi_{31}\cdot\cos\varphi_{32} \\ \sin\varphi_{32} \end{pmatrix},$$

out of which the φ_{31} angle can be expressed through the α, β, and γ

$$\operatorname{tg}\varphi_{31} = \frac{\cos\beta\cdot\sin\gamma}{\cos\alpha\cdot\cos\gamma - \sin\alpha\cdot\sin\beta\cdot\sin\gamma}$$

The spherical manipulator with three kinematic pair constraint equation can be expressed as the system of equations

$$\begin{cases} F_1 = \operatorname{tg}\varphi_{11} - \dfrac{\cos\beta\cdot\sin\alpha}{\cos\alpha\cdot\sin\alpha} = 0 \\ F_2 = \operatorname{tg}\varphi_{21} - \dfrac{\sin\beta}{\cos\alpha\cdot\cos\beta} = 0 \\ F_3 = \operatorname{tg}\varphi_{31} - \dfrac{\cos\beta\cdot\sin\gamma}{\cos\alpha\cdot\cos\gamma - \sin\alpha\cdot\sin\beta\cdot\sin\gamma} = 0 \end{cases}$$

Thus constraint equations for the input links and output links rotation angles are given.

6.3.4 Spherical Mechanism Velocity Problem

The velocity problem is solved by applying derivation method of spherical mechanism constraint equations.

Use the analytical method for velocity problem solution. It is based on the Jacobian matrix property study given in general as

$$C \cdot V = (-D) \cdot \omega_1 \tag{6.3}$$

$$C = \begin{pmatrix} \frac{\partial F_1}{\partial \alpha} & \frac{\partial F_1}{\partial \beta} & \frac{\partial F_1}{\partial \gamma} \\ \frac{\partial F_2}{\partial \alpha} & \frac{\partial F_2}{\partial \beta} & \frac{\partial F_2}{\partial \gamma} \\ \frac{\partial F_3}{\partial \alpha} & \frac{\partial F_3}{\partial \beta} & \frac{\partial F_3}{\partial \gamma} \end{pmatrix}; \; D = \begin{pmatrix} \frac{\partial F_1}{\partial \varphi_{11}} & 0 & 0 \\ 0 & \frac{\partial F_2}{\partial \varphi_{21}} & 0 \\ 0 & 0 & \frac{\partial F_3}{\partial \varphi_{31}} \end{pmatrix}; \; V = \begin{pmatrix} \dot{\alpha} \\ \dot{\beta} \\ \dot{\gamma} \end{pmatrix};$$

$$\omega_1 = \begin{pmatrix} \dot{\varphi}_{11} \\ \dot{\varphi}_{21} \\ \dot{\varphi}_{31} \end{pmatrix}$$

where C is the matrix containing partial derivatives of implicit function with respect to α, β, γ; D is the matrix containing partial derivatives of implicit function with respect to φ_{11}, φ_{21}, φ_{31} generalized coordinates; *V* is the output link velocities ω_1 are the generalized velocities in the spherical joints (input velocities).

We can resolve the input link velocity into α, β, γ and into *x, y, z* axis components. Velocity projections on the axes should be determined for the velocity problem solution. Assign $\dot{\alpha}, \dot{\beta}, \dot{\gamma}$ to the projections of the output link orientation angles evolution velocities V on the α, β, γ axes and ω_x, ω_y, ω_z.—on the *x, y, z* axes, respectively. We can write the relations of the projections as

$$\omega_x = \dot{\alpha} \cdot x_\alpha + \dot{\beta} \cdot x_\beta + \dot{\gamma} \cdot x_\gamma$$
$$\omega_y = \dot{\alpha} \cdot y_\alpha + \dot{\beta} \cdot y_\beta + \dot{\gamma} \cdot y_\gamma$$
$$\omega_z = \dot{\alpha} \cdot z_\alpha + \dot{\beta} \cdot z_\beta + \dot{\gamma} \cdot z_\gamma$$

$\begin{pmatrix} x_\alpha \\ y_\alpha \\ z_\alpha \end{pmatrix}$ are the kinematic pair coordinates of the fictitious chain which are equal to the product of matrix of rotation about the z axis by the coordinates of the 3-d kinematic pair.

The latter is in its initial position.

$$\begin{pmatrix} x_\alpha \\ y_\alpha \\ z_\alpha \end{pmatrix} = A \cdot \begin{pmatrix} 1 \\ 0 \\ 0 \end{pmatrix} = \begin{pmatrix} \cos\beta \cdot \cos\gamma \\ \cos\beta \cdot \sin\gamma \\ -\sin\beta \end{pmatrix}$$

$\begin{pmatrix} x_\beta \\ y_\beta \\ z_\beta \end{pmatrix}$ are the kinematic pair coordinates of the fictitious chain which are equal to the product of matrix of rotation about the z axis by the coordinates of the 3-d kinematic pair.

The layer is in initial position.

$$\begin{pmatrix} x_\beta \\ y_\beta \\ z_\beta \end{pmatrix} = A_3 \cdot \begin{pmatrix} 0 \\ 1 \\ 0 \end{pmatrix} = \begin{pmatrix} -\sin\gamma \\ \cos\gamma \\ 0 \end{pmatrix};$$

$\begin{pmatrix} x_\gamma \\ y_\gamma \\ z_\gamma \end{pmatrix}$ are the kinematic pair unit vector coordinates of the fictitious chain aligned on z axis.

Since z is a fixed axis we have the following x_α, y_β, z_γ coordinates:

$$\begin{pmatrix} x_\gamma \\ y_\gamma \\ z_\gamma \end{pmatrix} = \begin{pmatrix} 0 \\ 0 \\ 1 \end{pmatrix}$$

Substitute obtained values into the above mentioned equations

$$\begin{aligned} \omega_x &= \dot{\alpha} \cdot (\cos\beta \cdot \cos\gamma) + \dot{\beta} \cdot (-\sin\gamma) + \dot{\gamma} \cdot 0 \\ \omega_y &= \dot{\alpha} \cdot (\cos\beta \cdot \sin\gamma) + \dot{\beta} \cdot \cos\gamma + \dot{\gamma} \cdot 0 \\ \omega_z &= \dot{\alpha} \cdot (-\sin\beta) + \dot{\beta} \cdot 0 + \dot{\gamma} \cdot 1 \end{aligned}$$

Differentiate the following expressions:

$$\frac{\partial F_1}{\partial \alpha} = -\frac{\cos\beta \cdot \sin\alpha}{\cos\alpha^2};$$
$$\frac{\partial F_1}{\partial \beta} = \frac{\sin\beta}{\cos\alpha};$$
$$\frac{\partial F_1}{\partial \gamma} = 0;$$
$$\frac{\partial F_1}{\partial \varphi_{11}} = \mathrm{tg}\varphi_{11}^2 + 1;$$
$$\frac{\partial F_2}{\partial \alpha} = -\frac{\sin\beta \cdot \sin\alpha}{\cos\beta \cdot \cos\alpha^2};$$
$$\frac{\partial F_2}{\partial \beta} = -\frac{1}{\cos\alpha} - \frac{\sin\beta^2}{\cos\beta^2 \cdot \cos\alpha};$$
$$\frac{\partial F_2}{\partial \gamma} = 0;$$
$$\frac{\partial F_2}{\partial \varphi_{21}} = \mathrm{tg}\varphi_{21}^2 + 1;$$
$$\frac{\partial F_3}{\partial \alpha} = \frac{\sin\gamma \cdot \cos\beta \cdot (\cos\gamma \cdot \sin\alpha + \sin\gamma \cdot \cos\alpha \cdot \sin\beta)}{(\cos\gamma \cdot \cos\alpha - \sin\gamma \cdot \sin\beta \cdot \sin\alpha)^2};$$
$$\frac{\partial F_3}{\partial \beta} = \frac{\sin\gamma \cdot \sin\beta}{\cos\gamma \cdot \cos\alpha - \sin\gamma \cdot \sin\beta \cdot \sin\alpha} - \frac{\sin\gamma^2 \cdot \cos\beta^2 \sin\alpha}{(\cos\gamma \cdot \cos\alpha - \sin\gamma \cdot \sin\beta \cdot \sin\alpha)^2};$$
$$\frac{\partial F_3}{\partial \gamma} = \frac{\cos\gamma \cdot \cos\beta}{\cos\gamma \cdot \cos\alpha - \sin\gamma \cdot \sin\beta \cdot \sin\alpha} - \frac{\sin\gamma \cdot \cos\beta \cdot (\sin\gamma \cdot \cos\alpha + \cos\gamma \cdot \sin\beta \cdot \sin\alpha)}{(\cos\gamma \cdot \cos\alpha - \sin\gamma \cdot \sin\beta \cdot \sin\alpha)^2}$$
$$\frac{\partial F_3}{\partial \varphi_{31}} = \mathrm{tg}\varphi_{31}^2 + 1$$

6.3.4.1 Velocity Problem Solution by Applying the Theory of Screws

Consider the velocity inverse problem solution using the theory of screws.

The input links velocities should be obtained under the certain output links velocities. Note that the output link angular velocity is equal to the sum of one chain links joints angle velocities.

Write the equations for first kinematic chain

$$\begin{aligned} \omega_x &= \omega_{11} \cdot x_{11} + \omega_{12} \cdot x_{12} + \omega_{13} \cdot x_{13} \\ \omega_y &= \omega_{11} \cdot y_{11} + \omega_{12} \cdot y_{12} + \omega_{13} \cdot y_{13} \\ \omega_z &= \omega_{11} \cdot z_{11} + \omega_{12} \cdot z_{12} + \omega_{13} \cdot z_{13} \end{aligned} \tag{6.4}$$

where (x_{11}, y_{11}, z_{11}) are the first pair axis unit vector coordinates $\begin{pmatrix} x_{11} \\ y_{11} \\ z_{11} \end{pmatrix} = \begin{pmatrix} 1 \\ 0 \\ 0 \end{pmatrix}$;

(x_{12}, y_{12}, z_{12}) are the second pair axis unit vector coordinates, which are determined with the product of about x axis rotation matrix by the second pair axis unit vector coordinates given in its initial position

$$\begin{pmatrix} x_{12} \\ y_{12} \\ z_{12} \end{pmatrix} = \begin{pmatrix} 1 & 0 & 0 \\ 0 & \cos\varphi_{11} & -\sin\varphi_{11} \\ 0 & \sin\varphi_{11} & \cos\varphi_{11} \end{pmatrix} \cdot \begin{pmatrix} 0 \\ 1 \\ 0 \end{pmatrix} = \begin{pmatrix} 0 \\ \cos\varphi_{11} \\ \sin\varphi_{11} \end{pmatrix}$$

(x_{13}, y_{13}, z_{13}) are the third pair axis unit vector coordinates, which are determined with product of output link rotation matrix by the third pair axis unit vector coordinates given in initial position

$$\begin{pmatrix} x_{13} \\ y_{13} \\ z_{13} \end{pmatrix} = A \cdot \begin{pmatrix} 0 \\ 0 \\ 1 \end{pmatrix} = \begin{pmatrix} \sin\alpha \cdot \sin\gamma + \cos\alpha \cdot \cos\gamma \cdot \sin\beta \\ \cos\alpha \cdot \sin\beta \cdot \sin\gamma - \cos\gamma \cdot \sin\alpha \\ \cos\alpha \cdot \cos\beta \end{pmatrix}$$

Write the first kinematic chain velocities Eq. (6.4) in the following manner

$$\begin{pmatrix} \omega_x \\ \omega_y \\ \omega_z \end{pmatrix} = \begin{pmatrix} 1 & 0 & \sin\alpha \cdot \sin\gamma + \cos\alpha \cdot \cos\gamma \cdot \cos\beta \\ 0 & \cos\varphi_{11} & \cos\alpha \cdot \sin\beta \cdot \sin\gamma - \cos\gamma \cdot \sin\alpha \\ 0 & \sin\varphi_{11} & \cos\alpha \cdot \cos\beta \end{pmatrix} \cdot \begin{pmatrix} \omega_{11} \\ \omega_{12} \\ \omega_{13} \end{pmatrix} \tag{6.5}$$

Having solved given equation using the Kramer method, determine the first kinematic chain joints velocities.

Set up the second kinematic chain equations

$$\begin{aligned} \omega_x &= \omega_{21} \cdot x_{21} + \omega_{22} \cdot x_{22} + \omega_{23} \cdot x_{23} \\ \omega_y &= \omega_{21} \cdot y_{21} + \omega_{22} \cdot y_{22} + \omega_{23} \cdot y_{23} \\ \omega_z &= \omega_{21} \cdot z_{21} + \omega_{22} \cdot z_{22} + \omega_{23} \cdot z_{23} \end{aligned} \tag{6.6}$$

where (x_{21}, y_{21}, z_{21}) are the first pair axis unit vector coordinates of the second kinematic chain

$$\begin{pmatrix} x_{21} \\ y_{21} \\ z_{21} \end{pmatrix} = \begin{pmatrix} 0 \\ 1 \\ 0 \end{pmatrix};$$

(x_{22}, y_{22}, z_{22}) are the second pair axis unit vector coordinates, which are determined by the product of y axis rotation matrix by the second pair axis unit vector coordinates given in the initial position

$$\begin{pmatrix} x_{22} \\ y_{22} \\ z_{22} \end{pmatrix} = \begin{pmatrix} \cos\varphi_{21} & 0 & \sin\varphi_{21} \\ 0 & 1 & 0 \\ -\sin\varphi_{21} & 0 & \cos\varphi_{21} \end{pmatrix} \cdot \begin{pmatrix} 1 \\ 0 \\ 0 \end{pmatrix} = \begin{pmatrix} \cos\varphi_{21} \\ 0 \\ -\sin\varphi_{21} \end{pmatrix}$$

(x_{23}, y_{23}, z_{23}) are the third pair axis unit vector coordinates, which are determined by the product of output link rotation matrix by the third pair axis unit vector coordinates given in the initial position

$$\begin{pmatrix} x_{23} \\ y_{23} \\ z_{23} \end{pmatrix} = A \cdot \begin{pmatrix} 0 \\ 0 \\ 1 \end{pmatrix} = \begin{pmatrix} \sin\alpha \cdot \sin\gamma + \cos\alpha \cdot \cos\gamma \cdot \sin\beta \\ \cos\alpha \cdot \sin\beta \cdot \sin\gamma - \cos\gamma \cdot \sin\alpha \\ \cos\alpha \cdot \cos\beta \end{pmatrix}$$

Write the second kinematic chain velocities Eq. (6.6) in the following manner

$$\begin{pmatrix} \omega_x \\ \omega_y \\ \omega_z \end{pmatrix} = \begin{pmatrix} 0 & \cos\phi_{21} & \sin\alpha \cdot \sin\gamma + \cos\alpha \cdot \cos\gamma \cdot \cos\beta \\ 1 & 0 & \cos\alpha \cdot \sin\beta \cdot \sin\gamma - \cos\gamma \cdot \sin\alpha \\ 0 & -\sin\phi_{21} & \cos\alpha \cdot \cos\beta \end{pmatrix} \cdot \begin{pmatrix} \omega_{21} \\ \omega_{22} \\ \omega_{23} \end{pmatrix} \tag{6.7}$$

Set up the third kinematic chain equations:

$$\begin{aligned} \omega_x &= \omega_{31} \cdot x_{31} + \omega_{32} \cdot x_{32} + \omega_{33} \cdot x_{33} \\ \omega_y &= \omega_{31} \cdot y_{31} + \omega_{32} \cdot y_{32} + \omega_{33} \cdot y_{33} \\ \omega_z &= \omega_{31} \cdot z_{31} + \omega_{32} \cdot z_{32} + \omega_{33} \cdot z_{33} \end{aligned} \tag{6.8}$$

where (x_{31}, y_{31}, z_{31}) are the first pair axis unit vector coordinates of the third kinematic chain

$$\begin{pmatrix} x_{31} \\ y_{31} \\ z_{31} \end{pmatrix} = \begin{pmatrix} 0 \\ 0 \\ 1 \end{pmatrix};$$

(x_{32}, y_{32}, z_{32}) are the second pair axis unit vector coordinates, which are determined by the product of about z axis rotation matrix by the second pair axis unit vector coordinates given in the initial position

$$\begin{pmatrix} x_{32} \\ y_{32} \\ z_{32} \end{pmatrix} = \begin{pmatrix} \cos\varphi_{31} & -\sin\varphi_{31} & 0 \\ \sin\varphi_{31} & \cos\varphi_{31} & 0 \\ 0 & 0 & 1 \end{pmatrix} \cdot \begin{pmatrix} 1 \\ 0 \\ 0 \end{pmatrix} = \begin{pmatrix} \cos\varphi_{31} \\ \sin\varphi_{31} \\ 0 \end{pmatrix}$$

(x_{33}, y_{33}, z_{33}) are the third pair axis unit vector coordinates, which are determined by the product of output link rotation matrix by the third pair axis unit vector coordinates given in the initial position

$$\begin{pmatrix} x_{33} \\ y_{33} \\ z_{33} \end{pmatrix} = A \cdot \begin{pmatrix} 0 \\ 1 \\ 0 \end{pmatrix} = \begin{pmatrix} \cos\gamma \cdot \sin\beta \cdot \sin\alpha - \sin\gamma \cdot \cos\alpha \\ \cos\gamma \cdot \cos\alpha + \sin\gamma \cdot \sin\beta \cdot \sin\alpha \\ \cos\beta \cdot \sin\alpha \end{pmatrix}$$

Write the first kinematic chain velocities Eq. (6.8) in the following manner

$$\begin{pmatrix} \omega_x \\ \omega_y \\ \omega_z \end{pmatrix} = \begin{pmatrix} 0 & \cos\phi_{31} & \cos\gamma\cdot\sin\beta\cdot\sin\alpha - \sin\gamma\cdot\cos\alpha \\ 0 & \sin\phi_{31} & \cos\gamma\cdot\cos\alpha + \sin\gamma\cdot\sin\beta\cdot\sin\alpha \\ 1 & 0 & \cos\beta\cdot\sin\alpha \end{pmatrix} \cdot \begin{pmatrix} \omega_{31} \\ \omega_{32} \\ \omega_{33} \end{pmatrix} \quad (6.9)$$

Consider the velocities direct problem solution. Determine the output link velocities under the known input link velocities.

To solve this problem the wrench and tge twist [5] are assumed to be found.

The wrench $\boldsymbol{R_i}$ with (r_{ix}, r_{iy}, r_{iz}, r_{ix}^0, r_{iy}^0, r_{iz}^0) coordinates is reciprocal two unit vectors of the idle pairs e_{i2} and e_{i3} axes.

The R_i screw is equilibrated by the union screws i.e. back action forces in pairs corresponding to the unit vectors e_{i2} and e_{i3}.

The relative momentum mom($\boldsymbol{R}_i$, $\boldsymbol{\Omega}_i$) is the sum of the scalar products. Namely the scalar product of the first screw vector by the second screw momentum about one point and the scalar product of the second screw vector by the first screw momentum about the same point. Let us denote the output link twist as $\boldsymbol{\Omega}_\mathbf{i}$ with V_x, V_y, V_z, ω_x, ω_y, ω_z coordinates and denote the output link linear velocities as V_x, V_y, V_z.

The output link twist is the sum of chain links twists $\boldsymbol{\Omega}_i = \boldsymbol{\Omega}_{i1} + \boldsymbol{\Omega}_{i2} + \boldsymbol{\Omega}_{i3}$. Let us denote the first, second, and third pairs twists of the i-th link with (x_{i1}, y_{i1}, z_{i1}, x_{i1}^0, y_{i1}^0, z_{i1}^0) · ω_{i1}, (x_{i2}, y_{i2}, z_{i2}, x_{i2}^0, y_{i2}^0, z_{i2}^0) · ω_{i2}, (x_{i3}, y_{i3}, z_{i3}, x_{i3}^0, y_{i3}^0, z_{i3}^0) · ω_{i3} coordinates as $\boldsymbol{\Omega}_{i1}$, $\boldsymbol{\Omega}_{i2}$, $\boldsymbol{\Omega}_{i3}$.

Then two twists relative momentum is equal

mom($\boldsymbol{R}_i$, $\boldsymbol{\Omega}_\mathbf{i}$) = mom($\boldsymbol{R}_i$, $\boldsymbol{\Omega}_\mathbf{i1}$ + $\boldsymbol{R}_i$, $\boldsymbol{\Omega}_\mathbf{i2}$ + $\boldsymbol{R}_i$, $\boldsymbol{\Omega}_\mathbf{i3}$).

Since the wrenches is reciprocal to the nondriving pairs unit vectors, then the relative momentums mom($\boldsymbol{R}_i$, $\boldsymbol{\Omega}_\mathbf{i2}$) and mom($\boldsymbol{R}_i$, $\boldsymbol{\Omega}_\mathbf{i3}$) are equal to zero.

Therefore, we can write mom($\boldsymbol{R}_i$, $\boldsymbol{\Omega}_\mathbf{i2}$) = mom($\boldsymbol{R}_i$, $\boldsymbol{\Omega}_\mathbf{i1}$).

Substitute the twist and wrench coordinate values. We obtain the relative momentum equation:

$$\mathrm{mom}(\boldsymbol{R}_i, \Omega_\mathrm{i}) = \omega_x r_{ix}^0 + \omega_y r_{iy}^0 + \omega_z r_{iz}^0$$
$$\mathrm{mom}(\boldsymbol{R}_i, \Omega_\mathrm{i1}) = \omega_{i1}\cdot(x_{i1} r_{ix}^0 + y_{i1} r_{iy}^0 + z_{i1} r_{iz}^0)$$

where (x_{i1}, y_{i1}, z_{i1}) are the Plucker coordinates of the unit screws e_{i1} located in the axes of the first pairs; r_i^0 is the wrench momentum part with (r_{1x}^0, r_{1y}^0, r_{1z}^0) coordinates.

Set up the system of equations for three kinematic chains

$$\omega_x r_{1x}^0 + \omega_y r_{1y}^0 + \omega_z r_{1z}^0 = \omega_{11}\cdot(x_{11} r_{1x}^0 + y_{11} r_{1y}^0 + z_{11} r_{1z}^0)$$
$$\omega_x r_{2x}^0 + \omega_y r_{2y}^0 + \omega_z r_{2z}^0 = \omega_{21}\cdot(x_{21} r_{2x}^0 + y_{21} r_{2y}^0 + z_{21} r_{2z}^0)$$
$$\omega_x r_{3x}^0 + \omega_y r_{3y}^0 + \omega_z r_{3z}^0 = \omega_{31}\cdot(x_{31} r_{3x}^0 + y_{31} r_{3y}^0 + z_{31} r_{3z}^0)$$

where (x_{11}, y_{11}, z_{11}) are the first pair axis coordinates $\begin{pmatrix} x_{11} \\ y_{11} \\ z_{11} \end{pmatrix} = \begin{pmatrix} 1 \\ 0 \\ 0 \end{pmatrix}$;

(x_{12}, y_{12}, z_{12}) are the second pair axis coordinates that are equal to the product of the matrix specifying rotation about first axis by the second pair coordinates given in the initial position:

$$\begin{pmatrix} x_{12} \\ y_{12} \\ z_{12} \end{pmatrix} = \begin{pmatrix} 1 & 0 & 0 \\ 0 & \cos\alpha & -\sin\alpha \\ 0 & \sin\alpha & \cos\alpha \end{pmatrix} \cdot \begin{pmatrix} 0 \\ 1 \\ 0 \end{pmatrix} = \begin{pmatrix} 0 \\ \cos\alpha \\ \sin\alpha \end{pmatrix};$$

(x_{13}, y_{13}, z_{13}) are the third pair axis coordinates that are equal to the product of the matrix specifying output link rotation by the third pair coordinates given in the initial position:

$$\begin{pmatrix} x_{13} \\ y_{13} \\ z_{13} \end{pmatrix} = A \cdot \begin{pmatrix} 0 \\ 0 \\ 1 \end{pmatrix} = \begin{pmatrix} \sin\alpha \cdot \sin\gamma + \cos\alpha \cdot \cos\gamma \cdot \sin\beta \\ \cos\alpha \cdot \sin\beta \cdot \sin\gamma - \cos\gamma \cdot \sin\alpha \\ \cos\alpha \cdot \cos\beta \end{pmatrix}$$

In the case of first kinematic chain the wrench screw momentum part coordinates could be calculated in the following manner

$$r_1^0 = e_{12} \cdot e_{13} = \begin{bmatrix} \bar{i} & \bar{j} & \bar{k} \\ x_{12} & y_{12} & z_{12} \\ x_{13} & y_{13} & z_{13} \end{bmatrix}$$

Substituting unit vectors coordinate values we can obtain the power screw coordinate values.

We can determine the wrench r_2^0, r_3^0 momentum parts coordinates for the second and third chains

$$\begin{pmatrix} x_{21} \\ y_{21} \\ z_{21} \end{pmatrix} = \begin{pmatrix} 0 \\ 1 \\ 0 \end{pmatrix}; \begin{pmatrix} x_{22} \\ y_{22} \\ z_{22} \end{pmatrix} = \begin{pmatrix} \cos\beta & 0 & \sin\beta \\ 0 & 1 & 0 \\ -\sin\beta & 0 & \cos\beta \end{pmatrix} \cdot \begin{pmatrix} 1 \\ 0 \\ 0 \end{pmatrix} = \begin{pmatrix} \cos\beta \\ 0 \\ -\sin\beta \end{pmatrix};$$

$$\begin{pmatrix} x_{23} \\ y_{23} \\ z_{23} \end{pmatrix} = A \cdot \begin{pmatrix} 0 \\ 0 \\ 1 \end{pmatrix} = \begin{pmatrix} \sin\alpha \cdot \sin\gamma + \cos\alpha \cdot \cos\gamma \cdot \sin\beta \\ \cos\alpha \cdot \sin\beta \cdot \sin\gamma - \cos\gamma \cdot \sin\alpha \\ \cos\alpha \cdot \cos\beta \end{pmatrix};$$

$$\begin{pmatrix} x_{31} \\ y_{31} \\ z_{31} \end{pmatrix} = \begin{pmatrix} 0 \\ 0 \\ 1 \end{pmatrix}; \begin{pmatrix} x_{32} \\ y_{32} \\ z_{32} \end{pmatrix} = \begin{pmatrix} \cos\gamma & -\sin\gamma & 0 \\ \sin\gamma & \cos\gamma & 0 \\ 0 & 0 & 1 \end{pmatrix} \cdot \begin{pmatrix} 1 \\ 0 \\ 0 \end{pmatrix} = \begin{pmatrix} \cos\gamma \\ \sin\gamma \\ 0 \end{pmatrix}$$

$$\begin{pmatrix} x_{33} \\ y_{33} \\ z_{33} \end{pmatrix} = A \cdot \begin{pmatrix} 0 \\ 1 \\ 0 \end{pmatrix} = \begin{pmatrix} \cos\gamma \cdot \sin\beta \cdot \sin\alpha - \sin\gamma \cdot \cos\alpha \\ \cos\gamma \cdot \cos\alpha + \sin\gamma \cdot \sin\beta \cdot \sin\alpha \\ \cos\beta \cdot \sin\alpha \end{pmatrix},$$

Substituting unit vectors coordinate values we can obtain the power screw coordinate values for the second and third kinematic chains:

$$r_2^0 = e_{22} \cdot e_{23},$$
$$r_3^0 = e_{32} \cdot e_{33}$$

Therefore, the inverse and direct kinematic problems for translational-guiding mechanism were solved as well as the velocity problem was solved by the constraint equation differentiation method.

Direct and inverse kinematic problem were solved for spherical mechanism. The velocity problem was solved both by the constraint equation differentiation method and the screw method. The results of these two methods have coincided.

6.4 The Dynamic Properties Analysis of the PM with Kinematic Decoupling

The dynamic analysis turns out to be rather more general problem than kinematic analysis. The reason for this is that motor performance, law of control, feedback parameters, and etc. should be taken into account. The similar problem was set in a relatively small number of published papers [6–10].

First of all, the plain 3-DOF translational-guiding and spherical mechanisms were considered in the referenced papers. Dynamic motion was simulated under different motion laws, disagreement of dynamic model with real mechanism variables.

Different control procedures were used. Particularly the control procedure founded on the coordinate, velocity or acceleration error minimization was applied [8–12]. The mentioned algorithm is advantageous for the PMs according to the results of various numerical experiments.

Consider this control algorithm for the 6-DOF mechanism being studied. The decoupling concept should be used. Let us uncouple translational and rotational motions.

6.4.1 The 3-DOF Translation PM Movement

The solution of the specified problem is based on the dynamic model of movement of the mechanism being studied. With the given (target) law of the motion being applied, the inverse dynamic problem is solved and the required laws of actuating devices are determined. Motor control system calculates the required motor winding electric potential using relevant algorithm. Then the actual output link manipulator movements compared to the required ones are determined on the basis of the direct problem solution.

D'Alembert principle and principle of virtual work are used to set up the given mechanism movement equation. Give the elementary increments δq_1, δq_2, δq_3 of the input parameter and obtain the following motion equation about coordinate axes

$$\begin{aligned}
&m\ddot{x}\frac{\partial x}{\partial q_1}\delta q_1 + m\ddot{y}\frac{\partial y}{\partial q_1}\delta q_1 + m\ddot{z}\frac{\partial z}{\partial q_1}\delta q_1 + mg\frac{\partial z}{\partial q_1}\delta q_1 + m_1\ddot{q}_1\delta q_1 + P_1\delta q_1 = 0\\
&m\ddot{x}\frac{\partial x}{\partial q_2}\delta q_2 + m\ddot{y}\frac{\partial y}{\partial q_2}\delta q_2 + m\ddot{z}\frac{\partial z}{\partial q_2}\delta q_2 + mg\frac{\partial z}{\partial q_2}\delta q_2 + m_1\ddot{q}_2\delta q_2 + P_2\delta q_2 = 0\\
&m\ddot{x}\frac{\partial x}{\partial q_3}\delta q_3 + m\ddot{y}\frac{\partial y}{\partial q_3}\delta q_3 + m\ddot{z}\frac{\partial z}{\partial q_3}\delta q_3 + mg\frac{\partial z}{\partial q_3}\delta q_3 + m_1\ddot{q}_3\delta q_3 + P_3\delta q_3 = 0
\end{aligned} \tag{6.10}$$

Where m_1, m_2, m_3 are the masses of input links of the corresponding chains; m is the mass of output link; P_1, P_2, P_3 are the forces in actuating devices; $\ddot{x}, \ddot{y}, \ddot{z}$ are the accelerations of output link; $\ddot{q}_1, \ddot{q}_2, \ddot{q}_3$ are the accelerations in actuating devices.

For the coefficients $\frac{\partial x}{\partial q_1}, \frac{\partial x}{\partial q_2}, \frac{\partial x}{\partial q_3}$ to be determined it is necessary to consider the constraint equations written as the implicit coordinate system $F_i(x, y, z, q_i) = 0$ function.

In the given functions the variable q_i is presented as a function of x, y, z variables. Deriving each of these implicit functions with respect to x, y, z variables, we obtain the system of equations connecting velocities of the input and output links.

$$\frac{\partial F_i}{\partial x}\dot{x} + \frac{\partial F_i}{\partial y}\dot{y} + \frac{\partial F_i}{\partial z}\dot{z} + \frac{\partial F_i}{\partial q_i}\dot{q}_i = 0, \quad i = 1, 2, 3.$$

From the obtained expressions we get formulas to determine the unknown variable coefficients.

$$\frac{\partial x}{\partial q_i} = -\frac{\partial F_i}{\partial q_i}\Big/\frac{\partial F_i}{\partial x}, \frac{\partial y}{\partial q_i} = -\frac{\partial F_i}{\partial q_i}\Big/\frac{\partial F_i}{\partial y}, \frac{\partial z}{\partial q_i} = -\frac{\partial F_i}{\partial q_i}\Big/\frac{\partial F_i}{\partial z}$$

Suppose that the PM control algorithm constrained with the calculation of momentums and forces according to given law of motion must minimize acceleration, velocity, and coordinate errors. Earlier, in the paper [8] this approach was applied for the PM with two parallelograms in each chain Suppose that the required coordinate variation law $x_T(t)$, y_T (t), z_T (t) are known and the required velocities $\dot{x}_T(t), \dot{y}_T(t), \dot{z}_T(t)$ and accelerations $\ddot{x}_T(t), \ddot{y}_T(t), \ddot{z}_T(t)$ are also known respectively.

We are to determine the forces (momentums) when the errors in actuating devices are minimized with respect to coordinate

$$\Delta_1(t) = x_T(t) - x(t);\ \Delta_2(t) = y_T(t) - y(t);\ \Delta_3(t) = z_T(t) - z(t),$$

with respect to velocity:

$$\dot{\Delta}_1(t) = \dot{x}_T(t) - \dot{x}(t);\ \dot{\Delta}_2(t) = \dot{y}_T(t) - \dot{y}(t);\ \dot{\Delta}_3(t) = \dot{z}_T(t) - \dot{z}(t),$$

with respect to acceleration

$$\ddot{\Delta}_1(t) = \ddot{x}_T(t) - \ddot{x}(t);\ \ddot{\Delta}_2(t) = \ddot{y}_T(t) - \ddot{y}(t);\ \ddot{\Delta}_3(t) = \ddot{z}_T(t) - \ddot{z}(t).$$

The error law is known [8, 10] to correspond to the oscillatory link moving, when tolerance and error minimization with respect to coordinate, velocity, and acceleration are provided.

The output link acceleration laws in the feedback system are given by

$$\begin{aligned}\ddot{x} &= \ddot{x}_T + \gamma_1 \cdot (\dot{x}_T - \dot{x}) + \gamma_0 \cdot (x_T - x)\\ \ddot{y} &= \ddot{y}_T + \gamma_1 \cdot (\dot{y}_T - \dot{y}) + \gamma_0 \cdot (y_T - y)\\ \ddot{z} &= \ddot{z}_T + \gamma_1 \cdot (\dot{z}_T - \dot{z}) + \gamma_0 \cdot (z_T - z)\end{aligned} \tag{6.11}$$

where γ_0, γ_1 are the feedback coefficients.

These formulas determine the unknown laws of control forces and momentum generation, realizing movement along the designated trajectory.

In this research various initial conditions of manipulator motions, different feedback coefficients (γ_0, γ_1) were considered in order to obtain summary results.

Set coordinates values of the $A(x, y, z)$ point: x = 1 m, y = 1 m; z = 1 m. Link length L = 3 m.

Repeating differentiation of the Eq. (6.1) with respect to t variable we get the equations grouping together the input and output links:

$$\begin{aligned}
\ddot{q}_1 &= \ddot{x} - \frac{(2\cdot y\cdot\dot{y}+2\cdot z\cdot\dot{z})^2}{4\cdot(L^2-y^2-z^2)^{\frac{3}{2}}} - \frac{2\cdot\dot{y}^2+2\cdot\dot{z}^2+2\cdot y\cdot\ddot{y}+2\cdot z\cdot\ddot{z}}{2\cdot\sqrt{L^2-y^2-z^2}}\\
\ddot{q}_2 &= \ddot{y} - \frac{(2\cdot x\cdot\dot{x}+2\cdot z\cdot\dot{z})^2}{4\cdot(L^2-x^2-z^2)^{\frac{3}{2}}} - \frac{2\cdot\dot{x}^2+2\cdot\dot{z}^2+2\cdot x\cdot\ddot{x}+2\cdot z\cdot\ddot{z}}{2\cdot\sqrt{L^2-x^2-z^2}}\\
\ddot{q}_3 &= \ddot{z} - \frac{(2\cdot x\cdot\dot{x}+2\cdot y\cdot\dot{y})^2}{4\cdot(L^2-x^2-y^2)^{\frac{3}{2}}} - \frac{2\cdot\dot{x}^2+2\cdot\dot{y}^2+2\cdot x\cdot\ddot{x}+2\cdot y\cdot\ddot{y}}{2\cdot\sqrt{L^2-x^2-y^2}}
\end{aligned} \tag{6.12}$$

Absolute velocities from the Eq. (6.2) are expressed through the generalized velocities from the velocities equations.

Set the time constant $\tau \approx 0.018$ s. It characterizes the transient process time t and is equal to 0.076 s, in this case. Feedback factors: $\gamma_0 = 3086$, $\gamma_1 = 78.5$. Mass and weight force of the input links are omitted for simplicity.

In manipulator motion simulating under the known law the (6.1) equation is used. In this case forces P_i in the actuating devices must obey the equations which were obtained by substitution of $\ddot{x}, \ddot{y}, \ddot{z}$ accelerations from (6.12) equation in the motion Eq. (6.10) [8, 10].

$$P_i = m\cdot(\ddot{x}_T+\gamma_1\cdot(\dot{x}_T-\dot{x})+\gamma_0\cdot(x_T-x))\cdot\frac{\partial x}{\partial q_i} + m\cdot(\ddot{y}_T+\gamma_1\cdot(\dot{y}_T-\dot{y})+$$

$$\gamma_0\cdot(y_T-y))\cdot\frac{\partial y}{\partial q_i} + +m\cdot(\ddot{z}_T+\gamma_1\cdot(\dot{z}_T-\dot{z})+\gamma_0\cdot(z_T-z))\cdot\frac{\partial z}{\partial q_i} + mg\frac{\partial z}{\partial q_i} + m_i\ddot{q}_i,$$

where $i = 1, 2, 3$

Set the final control element motion law in view of

$$x_T(t) = 0.1\cdot\sin(\omega t);\; y_T(t) = 0.1\cdot\sin(\omega t);\; z_T(t) = 0.1\cdot\sin(\omega t). \tag{6.13}$$

Consider the motion at frequency $\omega = 40$ rad/s. The error differential equations become:

$$\begin{aligned}
\ddot{x} &= \ddot{x}_T + 78.5\cdot(\dot{x}_T-\dot{x}) + 3086\cdot(x_T-x)\\
\ddot{y} &= \ddot{y}_T + 78.5\cdot(\dot{y}_T-\dot{y}) + 3086\cdot(y_T-y)\\
\ddot{z} &= \ddot{z}_T + 78.5\cdot(\dot{z}_T-\dot{z}) + 3086\cdot(z_T-z)
\end{aligned}$$

The calculation resulted into error variation plots (Fig. 6.17) on different coordinates.

Analyzing the plots (Fig. 6.17) we can conclude: if the initial conditions are accepted and dynamic system of sinusoidal external action is imposed, the system takes stationary values within the limits of estimated unsteady process.

Here the appearance of negligible output link motion error is worth mentioning, the latter being the steady value too. Consider two cases. In the first one, the input

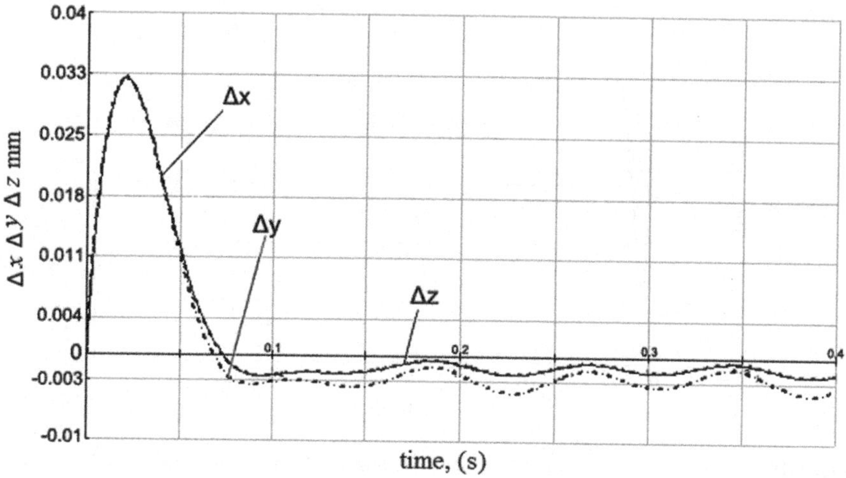

Fig. 6.17 Plot of error variation on the coordinate

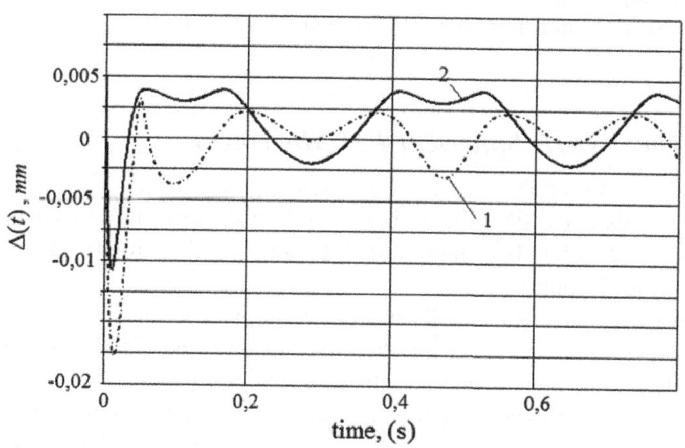

Fig. 6.18 Positioning error *1*—model input links and mechanism masses are different; *2*—model input links and mechanism masses are equal

link mass in a model is negligible i.e. mass is equal to zero. In the second one, the input link mass in a model is equal to real mass.

The deviation output link plots are presented on the Fig. 6.18. Since the difference of output link location error deviation is less than 5%, then we can omit the input link mass in the suggested algorithm.

Consider the case when the output link mass is not equal to the model output link mass.

Set the output link mass which differs from the model output link mass by 30%. Obviously, the obtained relations demonstrate the deviation of the robot output link

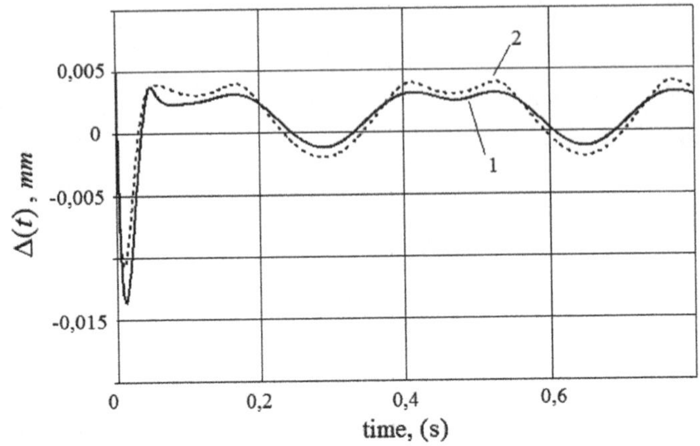

Fig. 6.19 Positioning error *1*—model output links and mechanism masses are equal; *2*—model output links and mechanism masses are different

mass from output link mass of the model within the limits of 30% which leads to position error deviation that does not exceed 3% (Fig. 6.19).

6.4.2 The 3-DOF Spherical PM Movement

Consider the spherical PM (Fig. 6.16) wherein each input link of chain is joined with a rotational motor. The output link is a platform, which rotates about three axis meeting in the *O* point.

Platform rotation angles α, β, γ are the output coordinates. The input links rotation angles φ_{11}, φ_{21}, φ_{31} of the first, second and third kinematic chains respectively are the generalized coordinates.

The moving coordinate system ξ, η, ζ assigns to the mechanism output link.

In the case of zero orientation angles (α = β = γ = 0) the directions of ξ, η, ζ axis are aligned with the directions of *x, y, z* axis, consequently.

The 3-DOF spherical PM motion equations are given by:

$$\begin{cases} J_\xi \cdot \ddot{\varphi}_\xi = M_1 \cdot \dfrac{\partial \varphi_{11}}{\partial \varphi_\xi} + M_2 \cdot \dfrac{\partial \varphi_{21}}{\partial \varphi_\xi} + M_3 \cdot \dfrac{\partial \varphi_{31}}{\partial \varphi_\xi} + \dot{\varphi}_\eta \cdot \dot{\varphi}_\zeta \cdot (J_\zeta - J_\eta) \\ J_\eta \cdot \ddot{\varphi}_\eta = M_1 \cdot \dfrac{\partial \varphi_{11}}{\partial \varphi_\eta} + M_2 \cdot \dfrac{\partial \varphi_{21}}{\partial \varphi_\eta} + M_3 \cdot \dfrac{\partial \varphi_{31}}{\partial \varphi_\eta} + \dot{\varphi}_\xi \cdot \dot{\varphi}_\zeta \cdot (J_\xi - J_\zeta) \\ J_\zeta \cdot \ddot{\varphi}_\zeta = M_1 \cdot \dfrac{\partial \varphi_{11}}{\partial \varphi_\zeta} + M_2 \cdot \dfrac{\partial \varphi_{21}}{\partial \varphi_\zeta} + M_3 \cdot \dfrac{\partial \varphi_{31}}{\partial \varphi_\zeta} + \dot{\varphi}_\xi \cdot \dot{\varphi}_\eta \cdot (J_\eta - J_\xi) \end{cases} \tag{6.14}$$

where J_ξ, J_η, J_ζ are the moments of inertia with respect to ξ, η, ζ axis; M_1, M_2, M_3 are the moments in actuation devices; $\frac{\partial \varphi_{ij}}{\partial \varphi_\eta}$ are the variable coefficients; $\dot{\varphi}_\xi, \ddot{\varphi}_\xi, \dot{\varphi}_\eta, \ddot{\varphi}_\eta, \dot{\varphi}_\zeta, \ddot{\varphi}_\zeta$ are the velocities and accelerations projections on the ξ, η, ζ axis.

Use the equations obtained in solving of velocity direct problem with screw method for variable coefficients $\frac{\partial \varphi_{ij}}{\partial \varphi_k}$ determination.

Equations system for three kinematic chains:

$$\begin{aligned}
\omega_\xi r^0_{1\xi} + \omega_\eta r^0_{1\eta} + \omega_\zeta r^0_{1\zeta} &= \omega_{11} \cdot (x_{11} r^0_{1x} + y_{11} r^0_{1y} + z_{11} r^0_{1z}) \\
\omega_\xi r^0_{2\xi} + \omega_\eta r^0_{2\eta} + \omega_\zeta r^0_{2\zeta} &= \omega_{21} \cdot (x_{21} r^0_{2x} + y_{21} r^0_{2y} + z_{21} r^0_{2z}) \\
\omega_\xi r^0_{3\xi} + \omega_\eta r^0_{3\eta} + \omega_\zeta r^0_{3\zeta} &= \omega_{31} \cdot (x_{31} r^0_{3x} + y_{31} r^0_{3y} + z_{31} r^0_{3z})
\end{aligned}$$

where ω_ζ, ω_η, ω_ζ are the input link angular velocities about ξ, η, ζ axis; $r_{i\xi}{}^0$, $r_{i\eta}{}^{00}$, $r_{i\zeta}{}^0$ are the momentum parts i-th force screw coordinates.

We can find the equation variable coefficients in the following manner:

$$\begin{aligned}
&\frac{\partial \varphi_{11}}{\partial \varphi_\xi} = \frac{\omega_{11}}{\omega_\xi} = \frac{r^0_{1\xi}}{r^0_{1x}}; \quad \frac{\partial \varphi_{11}}{\partial \varphi_\eta} = \frac{\omega_{11}}{\omega_\eta} = \frac{r^0_{1\eta}}{r^0_{1x}}; \quad \frac{\partial \varphi_{11}}{\partial \varphi_\zeta} = \frac{\omega_{11}}{\omega_\zeta} = \frac{r^0_{1\zeta}}{r^0_{1x}}; \\
&\frac{\partial \varphi_{21}}{\partial \varphi_\xi} = \frac{\omega_{21}}{\omega_\xi} = \frac{r^0_{2\xi}}{r^0_{2y}}; \quad \frac{\partial \varphi_{21}}{\partial \varphi_\eta} = \frac{\omega_{21}}{\omega_\eta} = \frac{r^0_{2\eta}}{r^0_{2y}}; \quad \frac{\partial \varphi_{21}}{\partial \varphi_\zeta} = \frac{\omega_{21}}{\omega_\zeta} = \frac{r^0_{2\zeta}}{r^0_{2y}}; \\
&\frac{\partial \varphi_{31}}{\partial \varphi_\xi} = \frac{\omega_{31}}{\omega_\xi} = \frac{r^0_{3\xi}}{r^0_{3z}}; \quad \frac{\partial \varphi_{31}}{\partial \varphi_\eta} = \frac{\omega_{31}}{\omega_\eta} = \frac{r^0_{3\eta}}{r^0_{3z}}; \quad \frac{\partial \varphi_{31}}{\partial \varphi_\zeta} = \frac{\omega_{31}}{\omega_\zeta} = \frac{r^0_{3\zeta}}{r^0_{3z}};
\end{aligned}$$

Set the required output link motion in a view of coordinate variation law: $\alpha_T(t), \beta_T(t), \gamma_T(t)$. Obtain the required velocities $\dot{\alpha}_T(t)$, $\dot{\beta}_T(t)$, $\dot{\gamma}_T(t)$ and accelerations $\ddot{\alpha}_T(t)$, $\ddot{\beta}_T(t)$, $\ddot{\gamma}_T(t)$ after differentiation. The problem of control is in errors minimization with respect to coordinate:

$$\Delta_1(t) = \alpha_T(t) - \alpha(t), \Delta_2(t) = \beta_T(t) - \beta(t), \Delta_3(t) = \gamma_T(t) - \gamma(t),$$

with respect to velocity:

$$\dot{\Delta}_1(t) = \dot{\alpha}_T(t) - \dot{\alpha}(t), \dot{\Delta}_2(t) = \dot{\beta}_T(t) - \dot{\beta}(t), \dot{\Delta}_3(t) = \dot{\gamma}_T(t) - \dot{\gamma}(t),$$

with respect to acceleration:

$$\ddot{\Delta}_1(t) = \ddot{\alpha}_T(t) - \ddot{\alpha}(t), \ddot{\Delta}_2(t) = \ddot{\beta}_T(t) - \ddot{\beta}(t), \ddot{\Delta}_3(t) = \ddot{\gamma}_T(t) - \ddot{\gamma}(t)$$

where $\alpha(t)$, $\beta(t)$, $\gamma(t)$ are the actual output link coordinate values.

Make the numerical simulation of PMs output link motion on given trajectory by applying optimal control of motion with the developed algorithm.

Set the damping coefficient $\zeta = \frac{\sqrt{2}}{2}$ and unsteady process time $t = 0.05$ s. In addition the reaction time is $\tau \approx 0.011$ s and feedback coefficients are $\gamma_0 = 7200$,

$\gamma_1 = 120$. The input links masses and weight force are omitted for simplicity. Here the motor parameters and maximum allowable generalized forces values are not taken into consideration.

The acceleration variation law is corresponded to the oscillatory link:

$$\begin{aligned}\ddot{\alpha} &= \ddot{\alpha}_T + \gamma_1 \cdot (\dot{\alpha}_T - \dot{\alpha}) + \gamma_0 \cdot (\alpha_T - \alpha);\\ \ddot{\beta} &= \ddot{\beta}_T + \gamma_1 \cdot (\dot{\beta}_T - \dot{\beta}) + \gamma_0 \cdot (\beta_T - \beta);\\ \ddot{\gamma} &= \ddot{\gamma}_T + \gamma_1 \cdot (\dot{\gamma}_T - \dot{\gamma}) + \gamma_0 \cdot (\gamma_T - \gamma).\end{aligned} \tag{6.15}$$

Express momentum M_i from Eq. (6.14). Substitute accelerations $\ddot{\alpha}, \ddot{\beta}, \ddot{\gamma}$ values from the Eq. (6.15) into obtained momentums expressions. Set the motion law of the final control element

$$\alpha_T(t) = 0.12 \cdot \sin(\omega t); \beta_T(t) = 0.12 \cdot \sin(\omega t); \gamma_T(t) = 0.15 \cdot sin(\omega t).$$

Then the accelerations differential equations become

$$\begin{aligned}\ddot{\alpha} &= \ddot{\alpha}_T + 120 \cdot (\dot{\alpha}_T - \dot{\alpha}) + 7200 \cdot (\alpha_T - \alpha)\\ \ddot{\beta} &= \ddot{\beta}_T + 120 \cdot (\dot{\beta}_T - \dot{\beta}) + 7200 \cdot (\beta_T - \beta)\\ \ddot{\gamma} &= \ddot{\gamma}_T + 120 \cdot (\dot{\gamma}_T - \dot{\gamma}) + 7200 \cdot (\gamma_T - \gamma)\end{aligned} \tag{6.16}$$

The position error variations and moments in actuating device plots in the case of feedback motion are shown in the Figs. 6.20 and 6.21, respectively.

Thus the 3-DOF spherical PM control algorithm ensuring errors on coordinate, velocity, and acceleration minimization is presented.

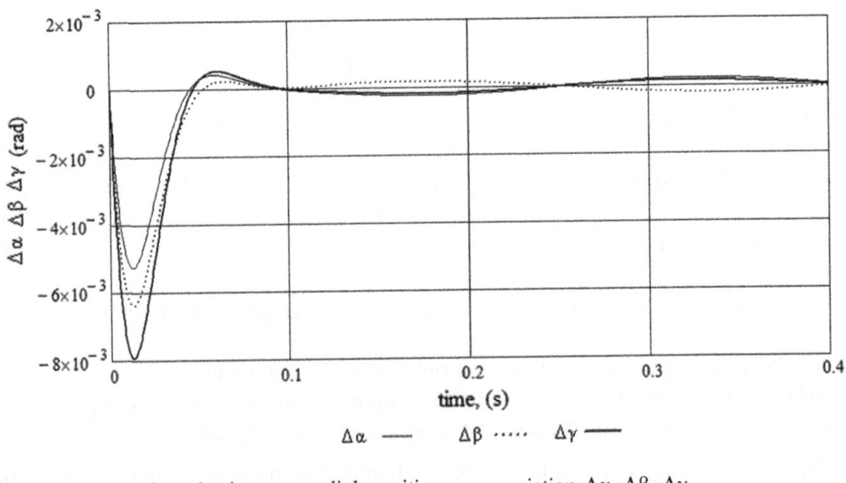

Fig. 6.20 Plot of mechanism output link position error variation Δα, Δβ, Δγ

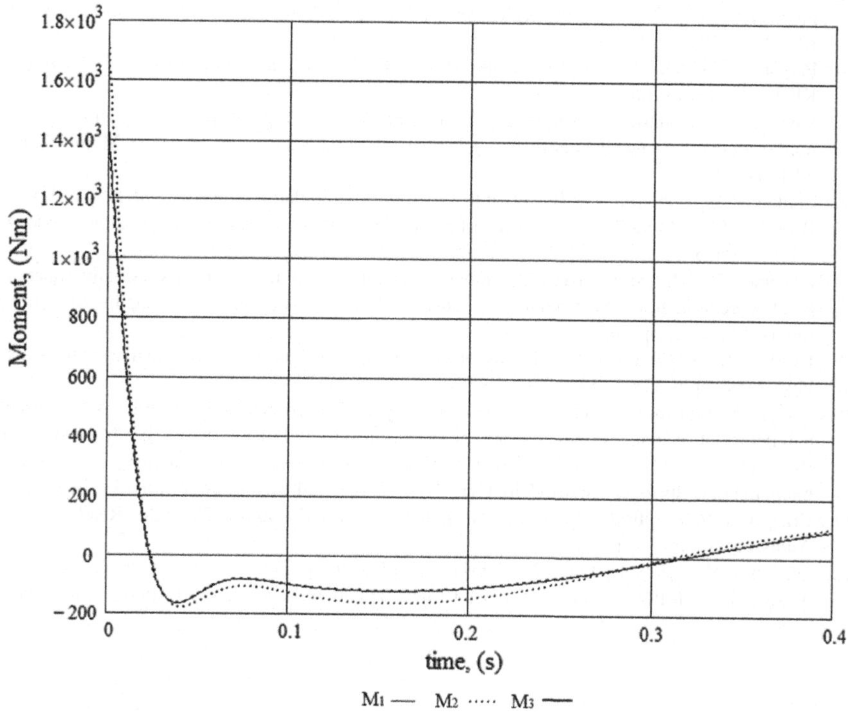

Fig. 6.21 Plot of the moments deviations in actuating device

6.5 Summary

The new 6-DOF, 5-DOF, 4-DOF PM were generated in this chapter. The mechanism possesses kinematic and dynamic decomposition. On the basis of the kinematic decomposition two mechanisms undergoing translation and rotation motions were obtained. Direct and inverse problems of each position and velocity for these mechanisms were solved. With the aid of dynamic decomposition the control algorithm founded on position, velocity, and acceleration errors minimization was tested.

References

1. Chablat D, Wenger P (2003, September) A new concept of modular parallel mechanism for machining applications. In: Proceedings of the 2003 IEEE international conference on robotics and automation (ICRA), Taipei, Taiwan, pp 14–19
2. Patent RU №2534706 Spatial mechanism of 4 DOF. Nosova N, Glazunov V, Palochkin S, Kheylo S (2013)

3. Patent RU №135283. Spatial mechanism of 5 DOF. Nosova N, Glazunov V, Palochkin S, Kheylo S (2013)
4. Patent RU №2536735. Spatial mechanism of 6 DOF. Nosova N, Glazunov V, Palochkin S, Kheylo S, Komisaruk L (2013)
5. Kheylo S, Glazunov V, Kulemkin Y, Ephros V (2013) Analysis of accelerations and nonlinear oscillations of spherical mechanism of parallel structure. J Mach Manuf Reliab 42:184–191
6. Zenkevich S, Ushenko A (2004) Control of robot manipulator. Moscow, MSTU, 480 p
7. Paul R (1976) Modelling, planning trajectory and control of robot manipulator. Moscow, Nauka, 104 p
8. Glazunov V, Kheylo S (2016) Dynamics and control of planar, translational, and spherical parallel manipulators. In: Dynamic balancing of mechanisms and synthesizing of parallel robots. Springer, pp 365–403
9. Krutko PD (1989) Inverse problems of control system dynamics: nonlinear model. Nauka Phys Math Publ 328 p (Moscow)
10. Kheylo S, Glazunov V (2014) Kinematics, dynamics, control and accuracy of spherical parallel robot. Advances on theory and practice of robots and manipulators. In: Proceedings of ROMANSY 2014 XX CISM-IFToMM symposium on theory and practice of robots and manipulators. Springer Cham Heidelberg New York Dordrecht London, pp 133–140
11. Craig JJ (1989) Introduction to robotics: mechanics and control, 2nd edn. Reading, MA: Addisson-Wesley
12. Glazunov V, Bykov R, Esina M (2004) Control of the parallel-structure mechanisms when passing through the singular positions. J Mach Manuf Reliab 2:73–79 (Allerton Press Inc.)

Chapter 7
Design of Decoupled Parallel Robots

T. Parikyan

Abstract Parallel robots often show superior accuracy, performance and load capacity alongside with a fairly simple control. Robot designs featuring kinematic and dynamic decoupling of motion, additionally improve the dynamic characteristics and further simplify the control algorithms. Parallel robots with 3D translatory motion of the end-effector (including micro-manipulator design variants) can be considered as mechatronic systems combining these advantages.

7.1 Introduction

7.1.1 Parallel Robots

From the mechanical point of view, the main technical requirements and quality criteria for the robot manipulators are:

- Accuracy—manifested by a level of positioning/orientation errors;
- Performance—max. attainable acceleration, as well as time for a sequence of repeatable actions;
- Load capacity—max. attainable force/moment of actuators/motors, as well as stiffness and durability of the structure;
- Simplicity of control system.

The invention of parallels robots, also known as platform-type robots (painting robot [13], tire testing machine [3] and the flight simulator being the first designs [16]), in most of the applications they have been designed for, showed their superiority—in the sense of the criteria mentioned above—versus traditional robots having in-series actuating mechanism. This, regardless the smaller workspace, opened new application fields for parallel robots not only in industry but also in

T. Parikyan (✉)
AVL LIST GmbH, Hans-List-Platz 1, 8020 Graz, Austria
e-mail: tigran.parikyan@avl.com

V. Arakelian (ed.), *Dynamic Decoupling of Robot Manipulators*, Mechanisms and Machine Science 56, https://doi.org/10.1007/978-3-319-74363-9_7

medicine (surgery), research, astronomy (telescopes), microscopy, in open space (vacuum), as well as within hazardous, dusty or dirty environment.

Differently to serial manipulators, the parallel ones feature structural stiffness just because of the fact of their frame-like design. To attain high performance and load capacity, one makes the frame as light and as stiff as possible: the best results are achieved here by using light and strong materials like carbon fiber reinforced plastic. At the same time, the actuators/motors are being placed on the base or near it, to prevent or to reduce the motion of these heaviest parts of the robot [1, 2].

Using *mini-* and *micro*-robots—in the sense of attainable accuracy, but also in the sense of the dimensions of the workspace compared to the dimensions of the robot structure itself—drastically improve the accuracy due to use of flexible gap-free joints and precise (e.g. piezoelectric) actuators.

Most of the parallel robot manipulators are neither kinematically nor dynamically decoupled: the typical examples are the well-known Delta robot [1, 2] and the Stewart-Gough platform [3, 16], as well as the less known first parallel robot—the Pollard painting robot [13]. In the latter one, for example, the structure and the geometry are the cause of the complicated kinematic equations [14], and though efforts have been taken to simplify the control algorithms of this kind of robots by appropriate selection of the structural and geometric parameters [12], these did not lead to any decoupling of motion.

7.1.2 Kinematic and Dynamic Decoupling

The dynamic characteristics and the controllability of robots can be improved by simplifying the trajectory planning and the equations of motion to solve, ideally by decoupling them, following the principle "one actuator—one DOF[1] of the end-effector".

The *kinematic decoupling* means that the motion from each linear or rotational actuator (or motor)—in the sense of velocities—is being transmitted to the respective motion of the end-effector[2] independently of the other actuators, i.e. the transmission matrix in the Eq. (7.1) is a diagonal one:

$$\begin{bmatrix} \dot{x} \\ \dot{y} \\ \dot{z} \end{bmatrix} = \begin{bmatrix} i_{1x} & 0 & 0 \\ 0 & i_{2y} & 0 \\ 0 & 0 & i_{3z} \end{bmatrix} \begin{bmatrix} \dot{q}_1 \\ \dot{q}_2 \\ \dot{q}_3 \end{bmatrix} \tag{7.1}$$

Should the transmission ratios on the diagonal of the matrix in Eq. (7.1) be variables (e.g. depending on the position of actuators q_i), we will speak of a *variable-ratio kinematic decoupling*.

[1]DOF = degree of freedom.

[2]Here robots with 3D translational motion of the end-effector are considered.

In case the transmission ratios on the diagonal of the matrix are constants, we will speak of a *constant-ratio kinematic decoupling*. In the most common and desired variant, we have a unity transmission matrix, where we can speak of *direct kinematic decoupling*:

$$\begin{bmatrix} \dot{x} \\ \dot{y} \\ \dot{z} \end{bmatrix} = \begin{bmatrix} 1 & 0 & 0 \\ 0 & 1 & 0 \\ 0 & 0 & 1 \end{bmatrix} \begin{bmatrix} \dot{q}_1 \\ \dot{q}_2 \\ \dot{q}_3 \end{bmatrix} \tag{7.2}$$

By saying *full kinematic decoupling* we will mean either constant-ratio kinematic decoupling or direct kinematic decoupling. Clearly, the kinematic control algorithms (or trajectory planning) are simpler in case of such full kinematic decoupling.

In case of full kinematic decoupling, the equations of motion (7.3) of an ideal mechanism[3] of robot relate the driving forces of the actuators F_i (or torques M_i of the motors) to the accelerations of the end-effector also through a diagonal mass matrix. This results in a *dynamic decoupling* between the actuators, i.e. the driving forces/moments are acting on the end-effector independently of one another:

$$\begin{bmatrix} m_{x1} & 0 & 0 \\ 0 & m_{y2} & 0 \\ 0 & 0 & m_{z3} \end{bmatrix} \begin{bmatrix} \ddot{x} \\ \ddot{y} \\ \ddot{z} \end{bmatrix} = \begin{bmatrix} F_1 \\ F_2 \\ F_3 \end{bmatrix} \tag{7.3}$$

If the diagonal coefficients of the mass matrix in (7.3) are variable (e.g. depending on the position of actuators q_i and their velocities $\dot{q}_i$), then we speak of a *variable-ratio dynamic decoupling*. If, instead, the diagonal mass matrix contains only constants, then we have a *constant-ratio* or *full dynamic decoupling*.

As we see later, full kinematic decoupling can lead to, but is not sufficient for, a full dynamic decoupling.

7.2 Parallel Robots with 3D Translational Motion of End-Effector

A class of parallel robots largely applicable in practice is represented by those with translational motion of the end-effector [1, 2]. This feature can also be successfully used to design decoupled parallel robots [4–11, 15].

The 3D translational motion of the end-effector of such robots is a result of simultaneous action of similar kinematic sub-chains connected in-parallel. Basically, there are two types of sub-chains used for this purpose: (a) those

[3]The bodies are considered rigid, connected by ideal kinematic pairs (joints) with no clearances. No stiffness, damping or friction are taken into account.

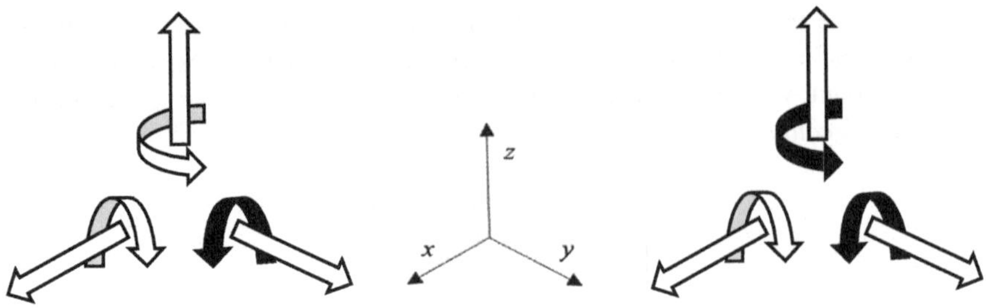

Fig. 7.1 Degrees of freedom (white arrows) and constraints (black arrows) imposed by a single kinematic sub-chain: **a** with one rotational constraint (left) and **b** with two rotational constraints

Table 7.1 Symbolic notation of kinematic pairs and chains

Symbolic notation	Description of kinematic pairs and chains
P	Prismatic kinematic pair (**P**—with linear actuator)
R	Revolute kinematic pair, or joint (**R**—with motor)
F	Planar ("Flat") kinematic pair
K	Cardan joint (compound kinematic pair $R^{\perp}R$ with crossing axes)
B	Bellow (compound flexible body and kinematic pair at the same time)
[CF]	"Cylinder-plane" contact: higher kinematic pair with axis of the cylinder parallel to the plane
$P^{\perp}F$	Two orthogonal kinematic pairs
$3^{\parallel}R$	Sequence of three revolute pairs with parallel axes
$3^{\perp}(\ldots)$	Three in-parallel mutually orthogonal kinematic sub-chains

imposing one rotational constraint, and (b) those with two such constraints (Fig. 7.1). Combining properly oriented sub-chains, one can impose a sufficient number of independent constraint conditions on rotations of the end-effector, so that it undergoes translational motions only. In case of in-parallel connected sub-chains, the sets (or sub-spaces) of constraints are united, while the sets (or sub-spaces) of the degrees of freedom are intersected [10].

In Table 7.1 you see the symbolic notation used throughout the chapter.

Below several original structures of robots with 3D translational motion of the end-effector are presented. Many of them can also be designed as precision positioners.

7.2.1 *Robot with Planar Pairs*

Figure 7.2 shows a parallel robot with 3D translatory motion of the end-effector. It consists of three base-mounted linear actuators, each connected with the

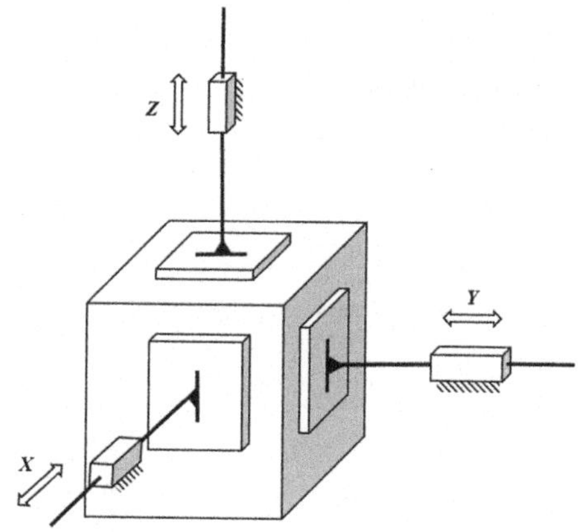

Fig. 7.2 Kinematically and dynamically fully decoupled $3^{\perp}(\mathbf{P}^{\perp}F)$ robot with base-mounted linear actuators and 3DOF 3D translatory motion of end-effector [15]

end-effector by a planar pair, here shown without a kinematic closure. The direction of motion of each linear actuator is perpendicular to the respective plane of contact with the end-effector. It has been first described in [15]. The structural formula of the mechanism of the robot is $3^{\perp}(\mathbf{P}^{\perp}F)$. Each $(\mathbf{P}^{\perp}F)$ sub-chain imposes two rotational constraints on the motion of the end-effector (Fig. 7.1b).

When a single actuator moves, it causes translatory motion of the end-effector in the same direction. There is no relative motion taking place in the contact of this actuator with the end-effector, while the other two planar contacts together form a prismatic pair, assuring its linear translatory motion. In case of simultaneous action, the motion of each actuator will be transmitted 1:1 to the end-effector along one of the three orthogonal directions, without any influence from the other actuators. Thus, the robot is fully kinematically decoupled.

The robot is also fully dynamically decoupled, as each actuator force causes acceleration in the same direction, depending on the constant mass of the end-effector and the moving part of the actuator itself (which is assumed to be constant, too), without any influence from the other ones. This also simplifies the control system of this robot.

Figure 7.3 shows a design variant of the $3^{\perp}(\mathbf{P}^{\perp}F)$ robot with actuators placed within a large-volume end-effector, that can be used to separate the inner volume from the outer environment [11].

Based on this design, a CNC-controlled lab prototype has been made (Fig. 7.4), where the base, the actuators and the balled planar pairs are shown within a cube-form light-weight plexiglas transparent end-effector.

7.2.2 Robot with Sub-chains Substituting Planar Pairs

Replacing the planar kinematic pairs F of the $3^{\perp}(\mathbf{P}^{\perp}F)$ robot described above, with three parallel rotational pairs (or joints) results in a structure $3^{\perp}(\mathbf{P}^{\parallel}3^{\parallel}R)$ shown in

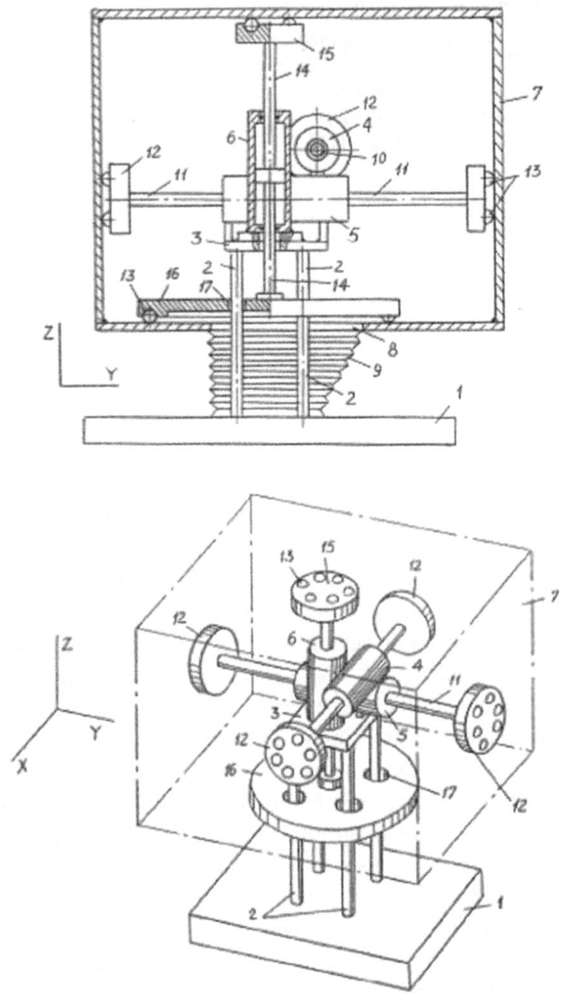

Fig. 7.3 Design variant of $3^{\perp}(\mathbf{P}^{\perp}\mathrm{F})$ robot with hermetically closed inner volume [11]

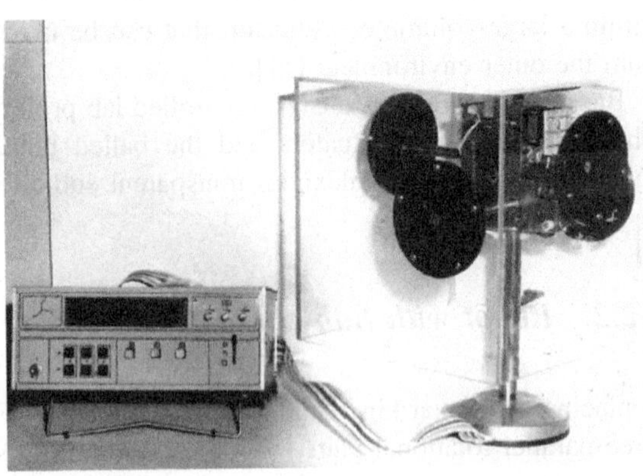

Fig. 7.4 The lab prototype of $3^{\perp}(\mathbf{P}^{\perp}\mathrm{F})$ robot made at Robotics Research Group of the State Engineering University of Armenia

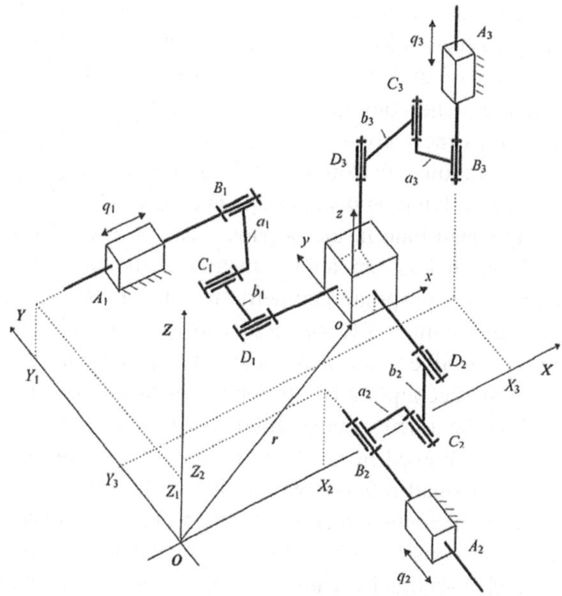

Fig. 7.5 Kinematically decoupled $3^{\perp}(\mathbf{P}^{\parallel}3^{\parallel}\mathrm{R})$ robot with three base-mounted linear actuators and 3D translatory motion of the end-effector

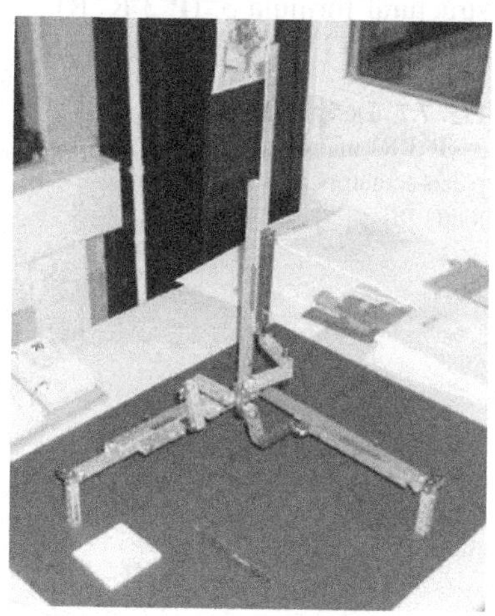

Fig. 7.6 A prototype of $3^{\perp}(\mathbf{P}^{\parallel}3^{\parallel}\mathrm{R})$ robot made by the author at the Institute of Microengineering of the Swiss Federal Institute of Technology in Lausanne (EPFL) [7]

Fig. 7.5 [5, 8]. Also here, each $(\mathbf{P}^{\parallel}3^{\parallel}\mathrm{R})$ sub-chain imposes two rotational constraints on the motion of the end-effector (Fig. 7.1b).

Figure 7.6 shows a prototype of the $3^{\perp}(\mathbf{P}^{\parallel}3^{\parallel}\mathrm{R})$ robot.

While the transmission of motion between the actuators and the end-effector here is 1:1 and remains constant (i.e. the robot is kinematically fully decoupled), the motion of each of the three $3^{\parallel}\mathrm{R}$ sub-chains is influenced by the linear actuators of the other two sub-chains [5].

It can be seen, that the robot is also dynamically decoupled, as each actuator causes acceleration of the end-effector in one respective direction only. However, the sum of the masses of the other two $3^{\parallel}$R sub-chains reduced to the current actuator—and consequently the value of the acceleration attained—depends on the position (linear coordinate) of the respective actuator. Consequently, in a general configuration, this robot manifests only a variable-ratio dynamic decoupling, but not a full dynamic decoupling. This makes the dynamic control of the robot more complicated than in the previous case of $3^{\perp}$(**P**$^{\perp}$F) robot having planar pairs.

Nevertheless, an approximate full dynamic decoupling can be achieved by choosing appropriate geometric parameters, namely in case when the links within each of the three $3^{\parallel}$R sub-chains are perpendicular to each other, and are parallel to the respective coordinate axes. With such a geometry, an instantaneous full dynamic decoupling of the robot mechanism is attained.

For the micro-robot variant of design (Fig. 7.7), the dynamic decoupling is always achieved in practice through nearly constant geometric configuration as a result of small motions within the mechanism.

The kinematic structure shown in Fig. 7.5, can be further improved by removing the over-constraints in order to prevent the structure from residual stresses and related inaccuracies caused by elastic deformations. This is carried out by introducing three perpendicular passive (compensating) joints (see Fig. 7.8), resulting in structural formula $3^{\perp}$(**P**$^{\parallel}3^{\parallel}R^{\perp}$R).

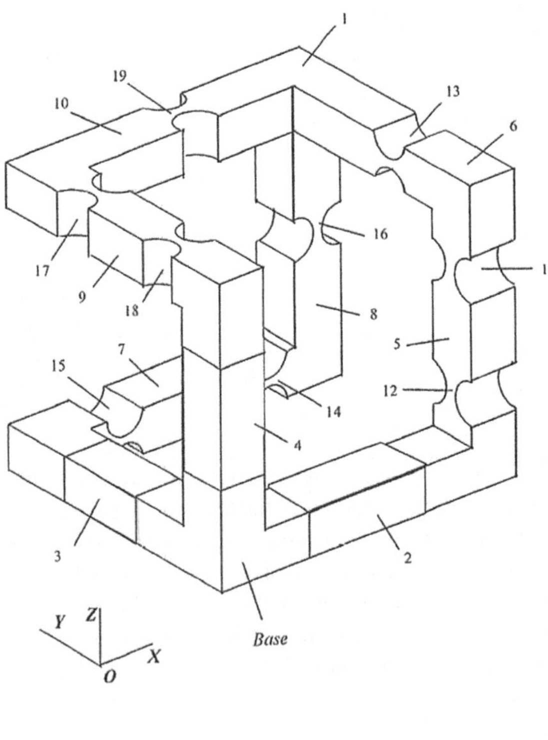

Fig. 7.7 Design variant of $3^{\perp}$(**P**$^{\parallel}3^{\parallel}$R) micro-robot with piezo-actuators and elastic joints [8]

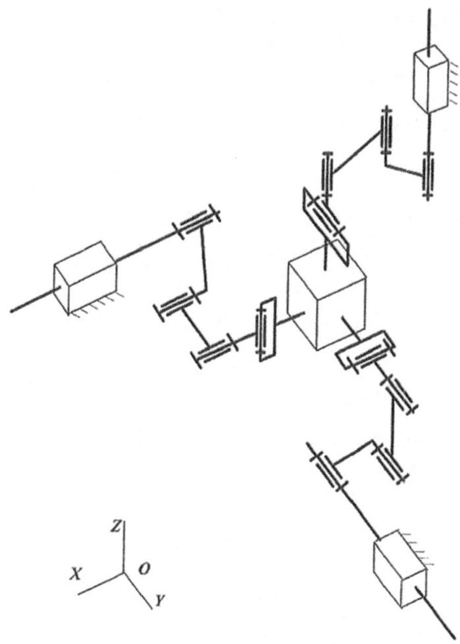

Fig. 7.8 Improved robot without over-constraints: structural formula $3^{\perp}(\mathbf{P}^{\parallel}3^{\parallel}\mathrm{R}^{\perp}\mathrm{R})$

In this case, each $(\mathbf{P}^{\parallel}3^{\parallel}\mathrm{R}^{\perp}\mathrm{R})$ sub-chain imposes only one rotational constraint on the motion of the end-effector (Fig. 7.1a).

Figure 7.9 shows the micro-robot design variant corresponding to the improved kinematic structure $3^{\perp}(\mathbf{P}^{\parallel}3^{\parallel}\mathrm{R}^{\perp}\mathrm{R})$, which can be used as a precision positioner.

7.2.3 Robot with Double Cardan Shafts

Figure 7.10 shows a parallel robot with 3D translatory motion of the end-effector. It consists of three base-mounted linear actuators, each connected with the end-effector by a double Cardan shaft. The axes of Cardan joints are perpendicular to the direction of motion of the linear actuators. It has been first described in [15].

The structural formula of the mechanism of the robot can be written as $3^{\perp}(\mathbf{P}^{\perp}\mathrm{K}^{\parallel}\mathrm{K})$. Each of the $(\mathbf{P}^{\perp}\mathrm{K}^{\parallel}\mathrm{K})$ sub-chains constraints the rotational motion of the end-effector about one of the orthogonal coordinate axes (Fig. 7.1a). All three sub-chains together constrain all possible rotations of the end-effector, therefore it can undergo translatory motions only.

When a single actuator moves, it causes circular translatory motion of the end-effector. To guide the end-effector along a straight line (e.g. a coordinate axis), all three actuators have to move. This means, that in general there is neither kinematic, nor dynamic decoupling for this kinematic scheme.

Fig. 7.9 Design variant of $3^{\perp}(\mathbf{P}^{\parallel}3^{\parallel}R^{\perp}R)$ micro-robot with piezo-actuators and elastic joints, without over-constraints [8]

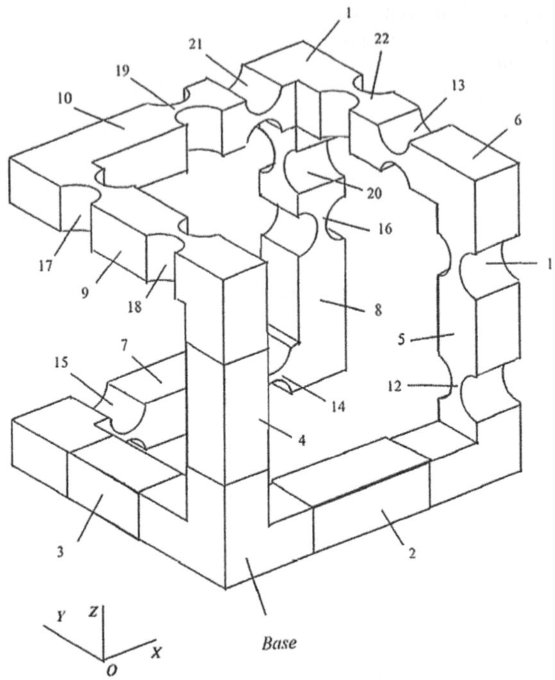

Fig. 7.10 Kinematic scheme of $3^{\perp}(\mathbf{P}^{\perp}K^{\parallel}K)$ robot with double Cardan shafts [15]

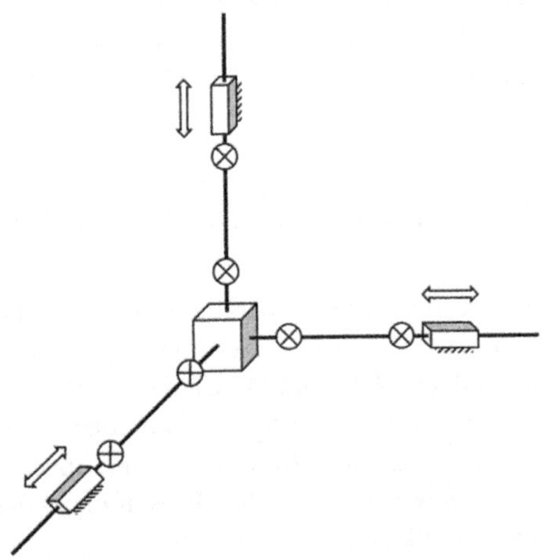

Figure 7.11 shows a design variant of the $3^{\perp}(\mathbf{P}^{\perp}K^{\parallel}K)$ micro-robot with piezo-electric linear actuators and flexible Cardan joints with restricted motion range [4].

For this design, both kinematic and dynamic full decoupling are always achieved in practice through a nearly constant geometric configuration, due to small motions of the actuators and in Cardan joints.

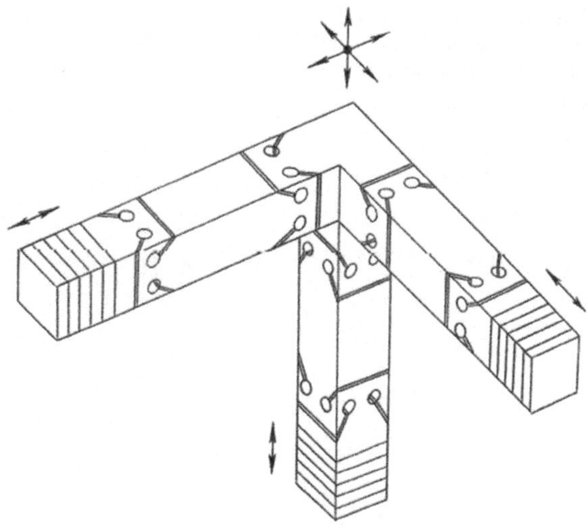

Fig. 7.11 Micro-robot $3^{\perp}(\mathbf{P}^{\perp}K^{\parallel}K)$ with double Cardan shafts and piezoelectric actuators [4]

The micro-robot described above can be used as a precision high-frequency positioner, which can be used e.g. in scanning devices.

7.2.4 *Robot with Bellows*

Figure 7.12 shows a parallel robot with 3D translatory motion of the end-effector. It consists of three mutually orthogonal bellows mounted with one end to the base, and with the other end—to the end-effector. On the end-effector, a grip or another tool can be attached, which has an additional, e.g. rotational DOF. It has been first described in [10].

The structural formula of the mechanism of the robot can be written as $3^{\perp}\mathbf{B}$. Each of the bellows constrains the rotational motion of the end-effector about one of the orthogonal coordinate axes (Fig. 7.1a).

Under relatively small motions ("mini-workspace"), translatory motion of the end-effector is assured due to the fact, that each of the bellows imposes a single torsional constraint, namely about its axis. Three such constraints about mutually orthogonal axes constrain any rotation, i.e. the end-effector can move only translatorily.

Each bellow is a kinematic sub-chain, a joint, and an actuator at the same time. Increasing or decreasing the pressure of the medium (gas or fluid) within one bellow causes its shortening or elongation. It causes also bending deformation of the other two bellows.

Such a design can be used where the hermetic isolation from the outer environment is needed. It is clear, however, that there is no kinematic or dynamic

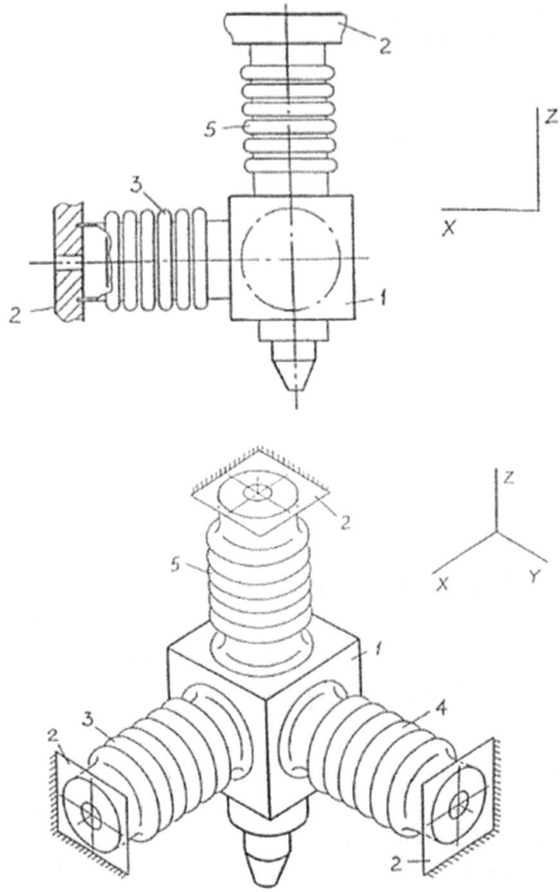

Fig. 7.12 Robot $3^{\perp}$**B** with three orthogonal bellows and 3D translatory motion of the end-effector

decoupling achieved by this robot, except for the closest vicinity to the nominal (non-deformed) state of all three bellows.

It is also not recommended to supply too high a pressure to the bellows and to put too high a load on the end-effector, as it can lead to larger deformations, caused by instability of the bellow shape, resulting in a significant deviation from translatory motion.

Based on this design, two lab prototypes have been made (Fig. 7.13), demonstrating the principles of motion control and enabling to assess the accuracy, stability and load capacity of the robot.

Figure 7.14 shows a possible design variant of the $3^{\perp}$**B** robot with symmetrically placed actuators for more stability, accuracy and load capacity.

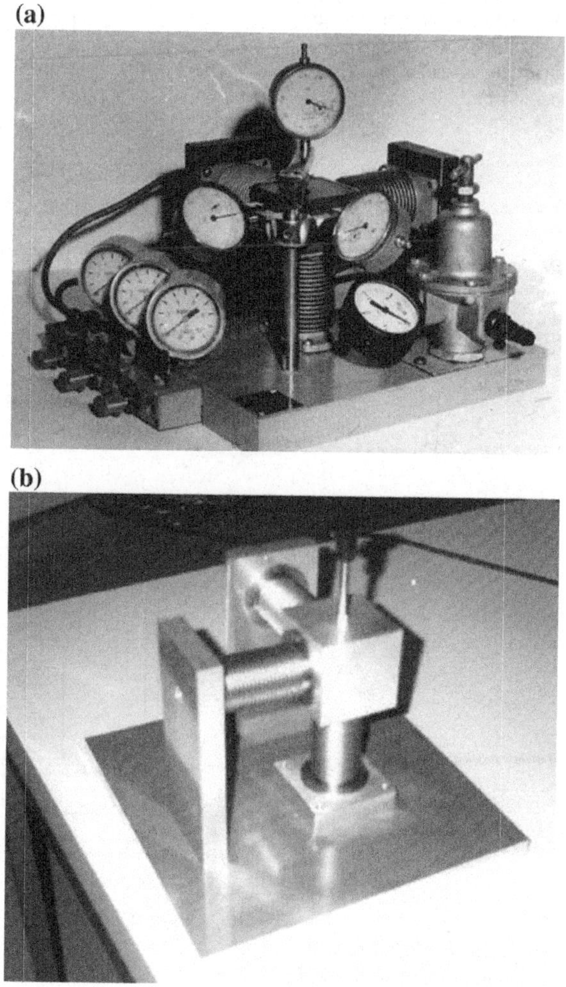

Fig. 7.13 Two lab prototypes of the $3^{\perp}\mathbf{B}$ robot: **a** on the left—made at the Department of Mechanical Engineering of the State Engineering University of Armenia, **b** on the right—made at the Institute of Microengineering of the Swiss Federal Institute of Technology in Lausanne (EPFL)

7.2.5 Robot with Bellows and Motors

Finally, introducing a motor between each bellow and the base results in a 6DOF robot shown in Fig. 7.15. Here, if very small motions are performed, so that the initial position and orientation of the bellows do not significantly change, one can practically attain both kinematic and dynamic full decoupling of all 6 components of motion:

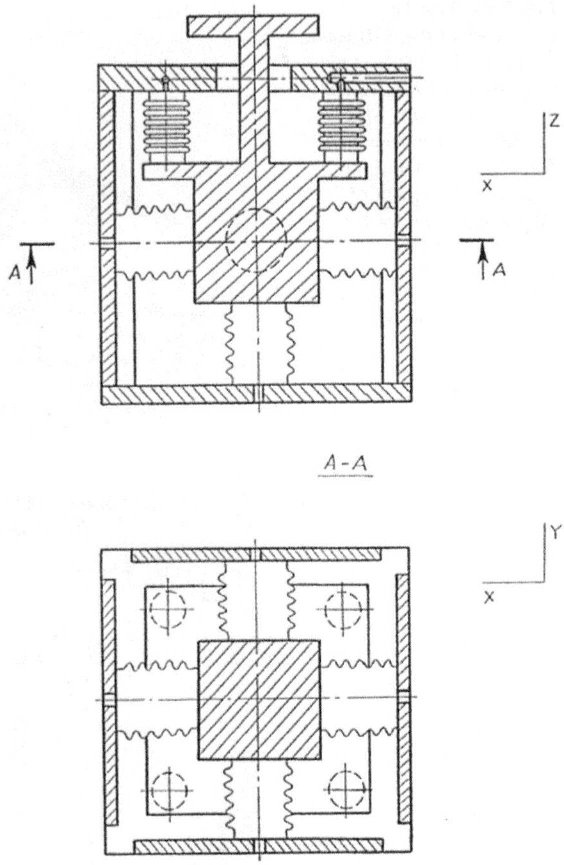

Fig. 7.14 Symmetric design variant of $3^{\perp}\mathbf{B}$ robot with improved mechanical parameters

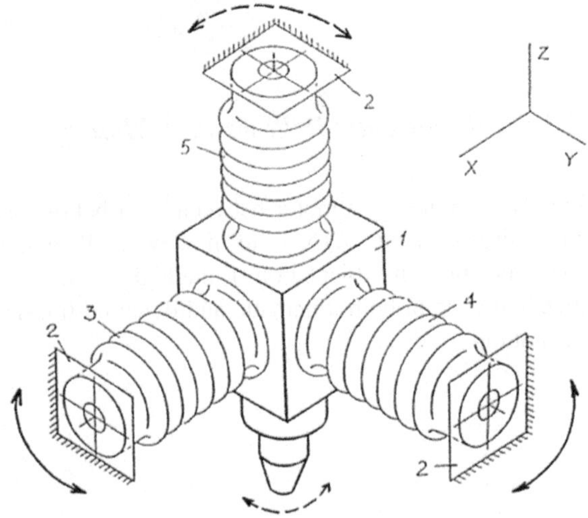

Fig. 7.15 Extended variant of the mini-robot with 6DOF $3^{\perp}(\mathbf{R}^{\parallel}\mathbf{B})$

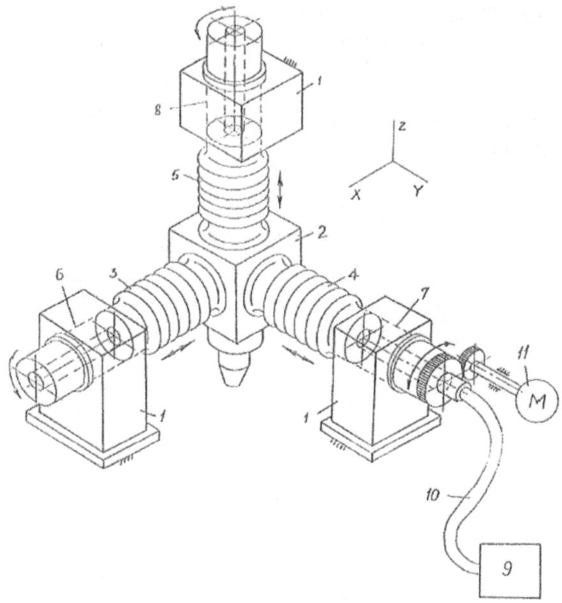

Fig. 7.16 Robot $3^{\perp}(\mathbf{R}^{\parallel}\mathbf{B})$ with 6DOF having three orthogonal bellows driven by motors and inflated by pumps [9]

$$\begin{bmatrix} \dot{x} \\ \dot{y} \\ \dot{z} \\ \dot{\varphi}_x \\ \dot{\varphi}_y \\ \dot{\varphi}_z \end{bmatrix} = \begin{bmatrix} 1 & & & & & \\ & 1 & & & & \\ & & 1 & & & \\ & & & 1 & & \\ & & & & 1 & \\ & & & & & 1 \end{bmatrix} \begin{bmatrix} \dot{q}_1 \\ \dot{q}_2 \\ \dot{q}_3 \\ \dot{q}_4 \\ \dot{q}_5 \\ \dot{q}_6 \end{bmatrix} \tag{7.4}$$

Figure 7.16 shows the patented design [9], with each sub-chain featuring a motor transmitting the rotation to the bellow through a gear, an a flexible hose to transmit the gas pressure into the inner volume of the bellow.

7.2.6 Robot with "Cylinder-Plane" Contacts

Figure 7.17 shows a parallel robot with 3D translatory motion of the end-effector [6]. It consists of three base-mounted motors, rotating eccentric cylinders, each connected with the end-effector by a double-sided "cylinder-plane" higher kinematic pair assuring kinematic closure. The axes of the motors are mutually orthogonal, the planes of the contacts are mutually orthogonal as well.

The structural formula of the mechanism can be written as $3^{\perp}(\mathbf{R}^{\parallel}[\mathrm{CF}])$.

When a single motor rotates, it causes rectilinear translatory motion of the end-effector in the direction perpendicular to the contact plane. With a constant rotation speed of the motor, the end-effector undergoes a sinusoidal motion. With

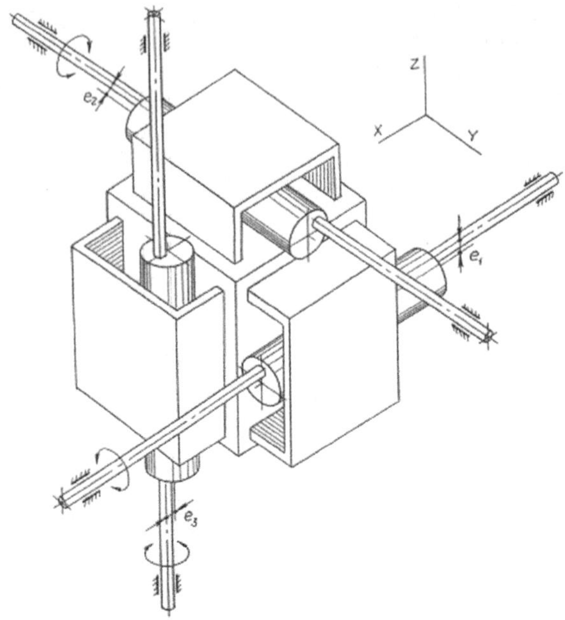

Fig. 7.17 Kinematically and dynamically variable-ratio decoupled robot $3^{\perp}(\mathbf{R}^{\parallel}[CF])$ with translatory motion of the end-effector and rotational actuators [6]

all three motors rotating with a constant speed, one can guide the end-effector along 3D harmonic curves of different types.

Although the motions are not transmitted 1:1 (they obey sinusoidal law), they are kinematically decoupled, i.e. each motor is causing a translational displacement along one coordinate axis only, and vice versa:

$$\begin{aligned} x &= e_1 \sin q_1 + x_0 \\ y &= e_2 \sin q_2 + y_0 \\ z &= e_3 \sin q_3 + z_0 \end{aligned} \tag{7.5}$$

or, after differentiation in time:

$$\begin{aligned} \dot{x} &= \dot{q}_1 e_1 \cos q_1 \\ \dot{y} &= \dot{q}_2 e_2 \cos q_2 \\ \dot{z} &= \dot{q}_3 e_3 \cos q_3 \end{aligned} \tag{7.6}$$

Re-writing the Eq. (7.6) in matrix form yields:

$$\begin{bmatrix} \dot{x} \\ \dot{y} \\ \dot{z} \end{bmatrix} = \begin{bmatrix} e_1 \cos q_1 & 0 & 0 \\ 0 & e_2 \cos q_2 & 0 \\ 0 & 0 & e_3 \cos q_3 \end{bmatrix} \begin{bmatrix} \dot{q}_1 \\ \dot{q}_2 \\ \dot{q}_3 \end{bmatrix} \tag{7.7}$$

Equation (7.7) shows that we have a clear case of a variable-ratio kinematic decoupling.

We note, that the robot is also dynamically decoupled, as each motor torque causes acceleration of the end-effector in one respective direction only. However, the moment of inertia reduced to each motor axis—and consequently the value of the acceleration of the end-effector, depend on the angular position of the respective motor. This means, that here we have a case of variable-ratio dynamic decoupling.

The described robot can be used as a high-speed positioner performing frequently repeated operations.

7.3 Summary

In this chapter, the principles of design of parallel robots with 3D translatory motion of end-effector are considered. Several original schemes and designs of parallel robots are shown, and it is discussed, under which conditions they become kinematically and dynamically decoupled. Two types of kinematic and dynamic decoupling are being distinguished: variable-ratio or constant-ratio (or full). It has also been shown, that full kinematic decoupling can—but not necessarily does—lead to full dynamic decoupling. On the other hand, due to small range of motion in actuators and flexible joints, micro-manipulator designs keep nearly the same geometric configuration, which in practice means both full kinematic and full dynamic decoupling of motion.

References

1. Clavel R (1991) Conception d'un robot parallèle rapide à 4 degrés de liberté. Thèse No 925, Ecole Polytechnique Fédérale de Lausanne, Lausanne
2. Clavel R (1985) Swiss Patent No. 672089 A5, Filed Dec 1985
3. Gough VE, Whitehall SG (1962) Universal tyre test machine. In: 9th International automobile technical congress. FISITA, IMechE, London, pp 117–137
4. Kochikian AV, Arutiunian MG, Parikian TF, Akopian AL, Ragulskis KM (1987) Mechanism of coordinate displacements. USSR Certificate of Authorship Patent No. 1814353, Filed 10 Aug 1987
5. Parikian TF (1996) Analysis and design of new kinematic structures for robots with high speed, accuracy and load characteristics. NATO Technical Report No. 460, Ministry of National Economy of Greece, Athens, 29 Mar 1996
6. Parikian TF (1988) Generator of 3-D mechanical vibrations. Patent of Russian Federation No. 1676668, Filed 28 Sept 1988
7. Parikian TF (1995) Kinematic analysis and design of parallel robots. Technical Report No. 95-02, Swiss Federal Institute of Technology, Lausanne
8. Parikyan T (1995) Manipulator. Armenian Patent No. 218, Filed 27 July 1995
9. Parikyan T (1996) Manipulator. Armenian Patent No. 322, Filed 26 Jan 1996
10. Parikian TF (1990) Development of methods for structural and kinematic analysis and synthesis of spatial parallel topology manipulation mechanisms. PhD thesis
11. Parikian TF, Yeghishian KM, Sarkissian YL (1987) Manipulator. Patent of Russian Federation No. 1463469, Filed 17 July 1987

12. Parikian T, Rey L, Clavel R (1995) Conception mécanique d'un robot parallèle permettant l'établissement de son modèle géométrique direct sous forme analytique, application au Pollard. In: Actes du 12e Congrès Français de Mécanique, vol 4. Strasbourg, France, pp 385–388
13. Pollard WLV (1942) Position-controlling apparatus. US Patent No. 2286371
14. Rey L (1994) Direct and inverse geometrical models of the Pollard's parallel robot. Internal Report, Swiss Federal Institute of Technology, Lausanne
15. Sarkissian YL, Parikian TF (1998) Principles of design of spatial translatory motion generating linkages. Mashinovedenie, No. 4, pp 12–20
16. Stewart D (1965–1966) A platform with six degrees of freedom. In: Proceedings of IMechE, London, vol180, Pt. l, No. 15, pp 371–386

The manufacturer's authorised representative in the EU is Springer Nature Customer Service Centre GmbH, Europaplatz 3, 69115 Heidelberg, Germany. If you have any concerns regarding our products, please contact ProductSafety@springernature.com

Printed and bound by CPI Group (UK) Ltd, Croydon, CR0 4YY
22/09/2026
02221064-0005